STEMD² Research & Development Group

Center on Disability Studies

College of Education

University of Hawaiʻi at Mānoa

http://stemd2.com/

Copyright ©2020 Center on Disability Studies, University of Hawaiʻi at Mānoa. All rights reserved. Printed in the United States of America. First published (2017) by the Center on Disability Studies, University of Hawaiʻi at Mānoa, Honolulu, Hawaiʻi. This book is based upon work supported by the Department of Education, Native Hawaiian Education Act Program under award #S362A140018. Any opinions, findings, and conclusions or recommendations expressed in this material are those of the authors and do not necessarily reflect the views of the United States Department of Education. For further information about this book and the Nā epapa Ka Hana project, please contact Dr. Kaveh Abhari at abhari@hawaii.edu.

ISBN: 978-0-9983142-2-8

Second release, 2020

Neʻepapa Ka Hana Eighth-Grade Mathematics Resources
Let's Play the ʻUkulele
Teacher's Guide

Project Directors	Kaveh Abhari
	Kelly Roberts
Content Developers	Robert G. Young
	Justin Toyofuku
Publication Designers	Katie Gao
	Robert G. Young
	Lauren Ho
	MyLan Tran
Content Contributors	Lock Lynch
	Luanna Peterson
	Holm Smidt
	Remy Pages
	Kimble McCann

Acknowledgments

We would like to thank Ruth Silberstein, Cody Kikuta, Katy Parsons, Crystal Yoo, Kelli Ching, Jacob Koseki, and Chase Cabana for advising on middle school mathematics education. Also, thank you to Keola Nakanishi and Mana Maoli, who were key motivators and supporters in the development of *Let's Play the ʻUkulele*.

Suggested Citation

Let's Play the ʻUkulele (2020). Neʻepapa Ka Hana Eighth-grade Mathematics Resources. STEMD^2 Book Series. STEMD^2 Research and Development Group. Center on Disability Studies, University of Hawaiʻi at Mānoa. Honolulu, Hawaiʻi.

Contents

Preface 7

Lesson Planning Structure 9
Introduction 9
Before the Lesson 11
During the Lesson 12
Wrap-up 14
Additional Notes 14
References 14

Common Core State Standards Alignment 15

Unit 1: Real Numbers, Exponents, and Scientific Notation 21
Activity 1.1 - The Meaning of Sound 22
Activity 1.2 - Octaves I 24
Activity 1.3 - Octaves II 27
Activity 1.4 - Notes of a Scale I 28
Activity 1.5 - Notes of a Scale II 31
Activity 1.6 - Notes of a Scale III 33
Activity 1.7 - Frequency Ranges 36

Unit 2: Proportional and Nonproportional Relationships and Functions .. 39
Activity 2.1 - The Speed of Sound 40
Activity 2.2 - Song Structure 43
Activity 2.3 - String Instruments I 47
Activity 2.4 - String Instruments II 50

Unit 3: Solving Equations and Systems of Equations 53
Activity 3.1 - Stage Setup I 54
Activity 3.2 - Stage Setup II 58

Unit 4: Transformational Geometry 61
Activity 4.1 - Semitones I 62

Activity 4.2 - Semitones II 64
Activity 4.3 - Chords 66
Activity 4.4 - 'Ukulele Design I 70
Activity 4.5 - 'Ukulele Design II 74

Unit 5: Measurement Geometry 77
Activity 5.1 - Sound reflections I 78
Activity 5.2 - Sound reflections II 81
Activity 5.3 - Reverberations 86
Activity 5.4 - Helmholtz Resonator 91

Unit 6: Statistics 95
Activity 6.1 - Reverberation Time 96
Activity 6.2 - Music Preferences I 100
Activity 6.3 - Music Preferences II 102

Preface

About the STEMD² Book Series

The STEMD² Book Series for eighth-grade mathematics was developed as part of a technology-enabled pedagogical approach (Neʻepapa Ka Hana model) for teaching mathematics in Hawaiʻi middle schools. This book series provides Hawaiʻi middle school teachers resources and training to incorporate problem-based learning, social learning and inclusive pedagogy through a culturally relevant mathematics curriculum. The series currently includes:

> **Let's Build a Canoe** – Student Activities and Teacher's Guide (Common Core aligned)
> **Let's Play the ʻUkulele** – Student Activities and Teacher's Guide (Common Core aligned)
> **Let's Go Fishing** – Student Activities and Teacher's Guide (SBAC aligned)
> **Let's Make Da Kine / E Hana Kākou** – Student Mini Projects and Teacher's Guide in English and ʻŌlelo Hawaiʻi (Skill development)

The printed and online resources produced by NKH through STEMD² are fully aligned with the Common Core Standards for Mathematical Practice and Content and the Smarter Balanced Assessments for mathematics. Based on the GO Math!® curriculum structure, the NKH STEMD² book series and social learning platform (`community.stemd2.com`) is flexible for teachers to implement partially or fully in their classrooms, as a tool to encourage students' interest and achievement in STEM subjects.

Funded by a three-year grant from the Department of Education's Native Hawaiian Education Act Program, Neʻepapa Ka Hana is a project of the STEMD² Research & Development Group in the Center on Disability Studies at the University of Hawaiʻi at Mānoa. More information about STEMD² is available online at `www.stemd2.com`.

Lesson Planning Structure

Introduction

Aloha kākou, e komo mai, and welcome to our **Neʻepapa Ka Hana (NKH)** guide for kumu! In this first chapter, we want to introduce to you the **Lesson Planning Structure** for developing, delivering, and reflecting upon your lessons. The planning suggestions presented provide guidelines for developing a range of meaningful lessons and memorable activities that truly engage your haumāna.

In the sections that follow, we will share a number of research-backed instructional resources and strategies that have been been adapted from the Mathematics Assessment Resource Service, University of Nottingham & UC Berkeley. For additional information and resources, we strongly encourage you to visit the Mathematics Assessment Project here: `http://map.mathshell.org/index.php`.

When reading through the guidelines, you will also find that many of the resources here complement the **GO Math!®** curriculum. In fact, they follow each of the **GO Math!®** lessons in their respective order. Please utilize your professional judgment to determine when and where to integrate these activities directly into your lessons. The alignment of the Common Core State Standards alignment and lessons the in this book are listed on Page 15.

For instance, these activities may serve as a culminating activity or Summative Assessment at the end of a lesson. It may also be appropriate to incorporate the **NKH Activity Set** as a Formative Assessment. Either way, the **NKH Activity Set** serves as a supplementary bridge to your mathematics curricula to enhance **Inclusive Classroom Pedagogy**, **Problem Based Learning**, and **Connectivism** (for the classes that utilize the website and online network capabilities).

The activities within the **NKH Activity Set** incorporate real-world problem solving tasks that reflect Hawaiian society, geography, and culture. The goal of these activities is for both kumu and haumāna to collaborate, communicate together, and think critically and creatively—and thus, deepen their understanding, application, and appreciation for mathematical thinking in the context of Hawaiian and island culture.

Formative Steps in the NKH Activity Set

In general, within each unit, activities can be used separately and in different orders. The cumulative activity should come after the different components of a module have been covered.

1. Before your Go Math!® lesson, students attempt the task individually.
2. After reviewing responses, you formulate questions for students to consider, geared toward improvements to their work.
3. At the start of the unit, play the introductory video to the NKH activity set. Then, let students think individually about the questions posed regarding their first attempt.
4. Next, students work in small groups to combine their thinking. Four students per group is often desired, as students could also work in pairs within the small group. The group works together to produce a collaborative solution to the given activity set (perhaps, in the form of a poster).
5. In the same small groups, students evaluate and comment on sample responses, identifying the strengths and weaknesses in these responses and comparing them with their own work.
6. In a whole-class discussion, students compare and evaluate the strategies they have seen and used.
7. In a follow-up lesson, students reflect for 10 minutes on their work and on what they have learned.

Materials Required

- **NKH Activity Set** or a **computer** with access to STEMD2 website (www.stemd2.com)
- A **Mini-Whiteboard** with **Marker** and **Eraser**—a quick way to visibly check individual understanding. This instructional strategy also enhances attention and participation.
- **Large sheets of paper** for small groups to create posters
- **Calculators and Graph Paper** for certain activities
- **Projector and Screen** to share the NKH introductory video to activity sets and to share students' sample responses.

Time Needed

- **Pre-Lesson Attempt:** 15-20 min.
- **NKH Activity Sets:** 50-60 min. (a full class period)
- **Follow-up and Reflection:** 5-10 min. at the beginning of each following class, to recap on preceding NKH Activity Sets.

Before the Lesson

Pre-Lesson Attempt

With the **Pre-Lesson Attempt** (15-20 min.), have your students informally complete this task in class or for homework before the actual **Formative Assessment** structured within the **NKH Activity Set**. This effort will provide you with an opportunity, in advance, to assess the work and to find out the kinds of difficulties that students may have. With this knowledge, you should then be able to target your interventions and strategies to help more effectively in the subsequent lesson.

Additionally, please remember to give each student a copy of the appropriate **NKH Activity Set** or give them the password needed to access the activities on the NKH website.

Examples of instructional prompts for individual student who are attempting any given **NKH Activity Set** include the following:

- Read the questions and try to answer them as carefully as you can.
- Show all your work, so that I can understand your reasoning.
- In addition to trying to solve the problem, I want you to check if you can present your work in a clear and organized fashion.

Students should work individually and without your assistance. Note that you may have to rearrange your students seating arrangements. However, this instructional strategy will allow you to have a more accurate and informative picture of your students' current understanding and levels of performance.

Assessing students' responses and giving feedback

After collecting the students' attempts at a **NKH Activity Set**, please take the time to create a few notes about what these samples of students' work reveal about their *current levels* of understanding and their individual and different approaches to problem-solving.

Scoring is not recommended during this phase.

It is also important to note that as a kumu, your feedback should summarize students' difficulties as a series of questions either by:

- Writing one or two questions on each student's work;
- Giving each student a printed version of your list of questions and highlight the questions that are more relevant to each individual student; or
- Selecting a few questions that will be of help to the majority of students and sharing them collectively with the whole class (either projected or written on the board) when returning students' initial attempts at the beginning of the NKH lesson.

When providing feedback to your haumāna, please refer to your own professional judgment and the respective needs of your individual instructional setting. That being said, certain common issues do arise across different classrooms and we recommend the following instructional prompts:

Common Issues	Feedback Examples
Student has difficulty getting started	• What do you know? • What do you need to find out?
Omits some given information when solving the problem	Write the given information in your own words.
Overlooks or misinterprets some contraints	• Can you organize ... in a systematic way? • What would make sense to try? Why? • How can you organize your work?
Makes incorrect assumptions	Will it always be the way you described?
Work is poorly presented	• Could someone unfamiliar with the task easily understand your work? • Have you explained how you arrived at your answer?
Provide little or no justification	How could you convince me that ... ?
Completes the task early	• How can you be sure that ... ? • What would happen if ... ? • Is there a way of describing all solutions?
Has technical difficulties	• Double check your work • Does it make sense? • Can you spot any mistakes?

A great deal of pedagogical creativity could take place in the pre-lesson task assessment/feedback. A highly recommended research reference is Hattie and Timperley (2007).

During the Lesson

When you are delivering the full **NKH Activity Set** as a lesson, please make sure that you have allotted approximately 50 minutes for proper lesson delivery and activity execution.

That being said, reviewing a student's first attempt at a **NKH Activity Set** should take approximately 5-10 minutes. This review is done individually and generally provided to the entire class, as a collective whole, by either projecting on a screen or writing on the board.

Remember to provide prompts to your haumāna continuously, for example:

- Recall what we were working on previously. What was the task about?
- I have had a look at your work. I would like you to think about the questions I wrote.
- On your own, carefully read through the questions I have written. I would like you to use questions to help you think about ways of improving your work.
- Use your mini–whiteboards to make a note of anything that you think will help improve your work.

It is imperative to have your students review their own work before working collaboratively in groups. When working independently, you may tailor your comments to individual students' to help them clarify what they are thinking.

Collaborative Small-Group Work
The first 5-10 minutes of these activities when students are divided into groups will be a time in which they will share and reflect upon their individual attempts at the prior, independent **NKH Activity Set**.

During this time, you may provide a few additional guiding prompts for your student groups:

- You each have your own individual solution to the task and have been thinking about how you might improve it, using the questions I have posed. Now, I want you to share your work with your partner(s). Take turns to explain how you did the task and how you think it could be improved.
- If explanations are unclear, ask questions until everyone in the group understands the individual solutions.

This leads to the next 15-20 minutes where students are jointly working on a common solution. The format of this collaborative activity can be elaborated upon a poster with a large sheet of paper. If you are using the NKH website, your joint solution can be posted onto the Forum. Notice that both of these formats will allow for sharing, discussing, and analyzing the different groups' approaches towards a solution. Of course, the use of your own professional judgment and any other format achieving this goal will suffice.

During this section, you should take note of the different approaches between groups, the change of directions, dialogue between groups, etc. This effort will help you guide the class discussion wrap-up.

Class management might/should be different from usual. This could be challenging. Remind your students that groups are like teams, and they will present and defend their team work. This may motivate students to keep on task while having fun and learning from each other. Who said that was impossible?

Additionally, you'll support and foster problem-solving skills by asking questions that help your haumāna clarify their thinking, while encouraging students to develop self-regulation as well as error detection skills.

Sharing, discussing, analyzing different approaches (10-20 minutes)
In this section of the **NKH Activity Set**, a whole-class discussion could follow the previous section. If posters were created, voluntary or randomly selected groups could share their strategies

that were developed towards a joint solution. It may be important to ask how the students' group solution differed from their individual solutions. If your students do not explicitly state their conclusions, you might ask how they checked their work. A conversation could also be initiated via the online Forum as well.

Another approach you might consider is sampling one or two groups' work to emphasize a particular aspect of their process. This is to help answer students' specific questions (either as a whole class or in each group) that pertain to their different approaches, procedures, methods, or errors where produced. For example, if a particular group previously arrived at a solution graphically, then have them do it algebraically. If errors were made but the reasoning was correct, ask how the errors could have been avoided.

Wrap-up (5-10 minutes)

As we wrap-up the final sections, a class discussion might conclude the **NKH Activity Set** by comparing the advantages and disadvantages of the approaches in the activity. Such discussion may center around shared difficulties or possible shortcuts that students could have developed either together or independently. It is important to recognize students' feelings and attitudes both during and after these activities.

Please note that the timing for these sections and activities range from 40 to 70 minutes but may vary from classroom to classroom depending on the nature of your needs within your particular instructional setting.

Additional Notes

The given lesson durations are only approximate. Please feel free to spend more or less time on these activities if it suits your classroom better. Furthermore, some of the very short activities can be skipped, but we discourage skipping the long ones. Lastly, the ↖ icon indicates where students can use the online learning platform at `www.stemd2.com`.

Please send us any comments, issues, technical or otherwise, you might have with the content, the format or the approach.

References

Hattie, J., & Timperley, H. (2007). The power of feedback. Review of educational research, 77(1), 81-112.

Mathematics Assessment Resource Service; University of Nottingham & UC Berkeley; `http://map.mathshell.org/index.php`.

Common Core State Standards Alignment

COMMON CORE STATE STANDARD	ʻUKULELE LESSON
The Number System	
Know that there are numbers that are not rational, and approximate them by rational numbers.	
8.NS.A.1 Know that numbers that are not rational are called irrational. Understand informally that every number has a decimal expansion; for rational numbers show that the decimal expansion repeats eventually, and convert a decimal expansion which repeats eventually into a rational number.	1.2, 1.5
8.NS.A.2 Use rational approximations of irrational numbers to compare the size of irrational numbers, locate them approximately on a number line diagram, and estimate the value of expressions (e.g., π^2). For example, by truncating the decimal expansion of $\sqrt{2}$, show that $\sqrt{2}$ is between 1 and 2, then between 1.4 and 1.5, and explain how to continue on to get better approximations.	1.6
Expressions and Equations	
Work with radicals and integer exponents.	
8.EE.A.1 Know and apply the properties of integer exponents to generate equivalent numerical expressions. For example, $3^2 \times 3^{-5} = 3^{-3} = 1/3^3 = 1/27$.	1.3, 1.4, 1.7
8.EE.A.2 Use square root and cube root symbols to represent solutions to equations of the form $x^2 = p$ and $x^3 = p$, where p is a positive rational number. Evaluate square roots of small perfect squares and cube roots of small perfect cubes. Know that $\sqrt{2}$ is irrational.	1.6

Continued on next page

COMMON CORE STATE STANDARD	'UKULELE LESSON
8.EE.A.3 Use numbers expressed in the form of a single digit times an integer power of 10 to estimate very large or very small quantities, and to express how many times as much one is than the other. For example, estimate the population of the United States as 3 times 10^8 and the population of the world as 7 times 10^9, and determine that the world population is more than 20 times larger.	1.7
8.EE.A.4 Perform operations with numbers expressed in scientific notation, including problems where both decimal and scientific notation are used. Use scientific notation and choose units of appropriate size for measurements of very large or very small quantities (e.g., use millimeters per year for seafloor spreading). Interpret scientific notation that has been generated by technology	1.7
Understand the connections between proportional relationships, lines, and linear equations.	
8.EE.B.5 Graph proportional relationships, interpreting the unit rate as the slope of the graph. Compare two different proportional relationships represented in different ways. For example, compare a distance-time graph to a distance-time equation to determine which of two moving objects has greater speed.	2.1, 2.2
8.EE.B.6 Use similar triangles to explain why the slope m is the same between any two distinct points on a non-vertical line in the coordinate plane; derive the equation $y = mx$ for a line through the origin and the equation $y = mx + b$ for a line intercepting the vertical axis at b.	
Analyze and solve linear equations and pairs of simultaneous linear equations.	
8.EE.C.7 Solve linear equations in one variable.	2.1, 2.2, 3.2, 5.3
8.EE.C.7.a Give examples of linear equations in one variable with one solution, infinitely many solutions, or no solutions. Show which of these possibilities is the case by successively transforming the given equation into simpler forms, until an equivalent equation of the form $x = a$, $a = a$, or $a = b$ results (where a and b are different numbers).	3.2, 5.3
8.EE.C.7.b Solve linear equations with rational number coefficients, including equations whose solutions require expanding expressions using the distributive property and collecting like terms.	2.1, 2.2

Continued on next page

COMMON CORE STATE STANDARD	'UKULELE LESSON
8.EE.C.8 Analyze and solve pairs of simultaneous linear equations.	3.1, 3.2
8.EE.C.8.a Understand that solutions to a system of two linear equations in two variables correspond to points of intersection of their graphs, because points of intersection satisfy both equations simultaneously.	3.1, 3.2
8.EE.C.8.b Solve systems of two linear equations in two variables algebraically, and estimate solutions by graphing the equations. Solve simple cases by inspection. For example, 3x + 2y = 5 and 3x + 2y = 6 have no solution because 3x + 2y cannot simultaneously be 5 and 6.	3.1, 3.2
8.EE.C.8.c Solve real-world and mathematical problems leading to two linear equations in two variables. For example, given coordinates for two pairs of points, determine whether the line through the first pair of points intersects the line through the second pair.	3.1
Functions	
Define, evaluate, and compare functions.	
8.F.A.1 Understand that a function is a rule that assigns to each input exactly one output. The graph of a function is the set of ordered pairs consisting of an input and the corresponding output.	2.4
8.F.A.2 Compare properties of two functions each represented in a different way (algebraically, graphically, numerically in tables, or by verbal descriptions).For example, given a linear function represented by a table of values and a linear function represented by an algebraic expression, determine which function has the greater rate of change.	2.2, 3.2
8.F.A.3 Interpret the equation y = mx + b as defining a linear function, whose graph is a straight line; give examples of functions that are not linear. For example, the function $A = s^2$ giving the area of a square as a function of its side length is not linear because its graph contains the points (1,1), (2,4) and (3,9), which are not on a straight line.	2.1, 2.4, 3.1, 3.2
Use functions to model relationships between quantities.	

Continued on next page

COMMON CORE STATE STANDARD	'UKULELE LESSON
8.F.B.4 Construct a function to model a linear relationship between two quantities. Determine the rate of change and initial value of the function from a description of a relationship or from two (x, y) values, including reading these from a table or from a graph. Interpret the rate of change and initial value of a linear function in terms of the situation it models, and in terms of its graph or a table of values.	2.1, 2.2, 3.1
8.F.B.5 Describe qualitatively the functional relationship between two quantities by analyzing a graph (e.g., where the function is increasing or decreasing, linear or nonlinear). Sketch a graph that exhibits the qualitative features of a function that has been described verbally.	2.2, 2.3, 3.2
Geometry	
Understand congruence and similarity using physical models, transparencies, or geometry software.	
8.G.A.1 Verify experimentally the properties of rotations, reflections, and translations:	4.2, 4.3, 4.5, 5.1, 5.2
8.G.A.1.a Lines are taken to lines, and line segments to line segments of the same length.	4.3
8.G.A.1.b Angles are taken to angles of the same measure.	4.3, 5.1, 5.2
8.G.A.1.c Parallel lines are taken to parallel lines.	4.3
8.G.A.2 Understand that a two-dimensional figure is congruent to another if the second can be obtained from the first by a sequence of rotations, reflections, and translations; given two congruent figures, describe a sequence that exhibits the congruence between them.	4.3, 4.5
8.G.A.3 Describe the effect of dilations, translations, rotations, and reflections on two-dimensional figures using coordinates.	4.2, 4.3, 4.4, 4.5
8.G.A.4 Understand that a two-dimensional figure is similar to another if the second can be obtained from the first by a sequence of rotations, reflections, translations, and dilations; given two similar two-dimensional figures, describe a sequence that exhibits the similarity between them.	4.4, 4.5

Continued on next page

COMMON CORE STATE STANDARD	**'UKULELE LESSON**
8.G.A.5 Use informal arguments to establish facts about the angle sum and exterior angle of triangles, about the angles created when parallel lines are cut by a transversal, and the angle-angle criterion for similarity of triangles. For example, arrange three copies of the same triangle so that the sum of the three angles appears to form a line, and give an argument in terms of transversals why this is so.	5.1, 5.2
Understand and apply the Pythagorean Theorem.	
8.G.B.6 Explain a proof of the Pythagorean Theorem and its converse.	
8.G.B.7 Apply the Pythagorean Theorem to determine unknown side lengths in right triangles in real-world and mathematical problems in two and three dimensions.	5.3
8.G.B.8 Apply the Pythagorean Theorem to find the distance between two points in a coordinate system.	5.3
Solve real-world and mathematical problems involving volume of cylinders, cones, and spheres.	
8.G.C.9 Know the formulas for the volumes of cones, cylinders, and spheres and use them to solve real-world and mathematical problems.	5.4
Statistic and Probability	
Investigate patterns of association in bivariate data.	
8.SP.A.1 Construct and interpret scatter plots for bivariate measurement data to investigate patterns of association between two quantities. Describe patterns such as clustering, outliers, positive or negative association, linear association, and nonlinear association.	6.1
8.SP.A.2 Know that straight lines are widely used to model relationships between two quantitative variables. For scatter plots that suggest a linear association, informally fit a straight line, and informally assess the model fit by judging the closeness of the data points to the line.	6.1

Continued on next page

COMMON CORE STATE STANDARD	'UKULELE LESSON
8.SP.A.3 Use the equation of a linear model to solve problems in the context of bivariate measurement data, interpreting the slope and intercept. For example, in a linear model for a biology experiment, interpret a slope of 1.5 cm/hr as meaning that an additional hour of sunlight each day is associated with an additional 1.5 cm in mature plant height.	6.1
8.SP.A.4 Understand that patterns of association can also be seen in bivariate categorical data by displaying frequencies and relative frequencies in a two-way table. Construct and interpret a two-way table summarizing data on two categorical variables collected from the same subjects. Use relative frequencies calculated for rows or columns to describe possible association between the two variables. For example, collect data from students in your class on whether or not they have a curfew on school nights and whether or not they have assigned chores at home. Is there evidence that those who have a curfew also tend to have chores?	6.2, 6.3

Unit 1: Real Numbers, Exponents, and Scientific Notation

Activity 1.1 - The Meaning of Sound

When a rock falls into water, the water gets pushed and pulled around, creating little waves that spread out. This also happens in air, which is how we hear *sound*. When something like a 'ukulele string moves (vibrates), it squishes and stretches the air around it. This squishing and stretching of air then spreads around as a *sound wave*. When the air squishes and stretches quickly and the wave reach our ears just right, we call this *sound*.

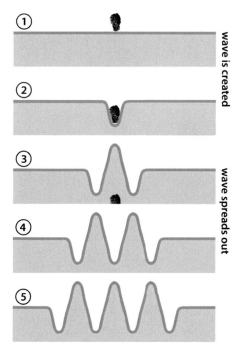

The *frequency* of a sound describes how often the air squishes and stretches in one second, and its unit is called a hertz (Hz). Hypothetically, if you could hear a sound of 1 Hz, then your ears are noticing that the air is squished and stretched 1 time each second. If you could hear 9 kHz (kilohertz), then your ears are noticing that the air is squished and stretched 9,000 times per second. In music, some specific frequencies are called *notes*, and we give them names like B, C flat, and F sharp.

You might have heard people talk about sounds that are "high pitch" or "low pitch." Like frequency, *pitch* describes how many times a sound wave is squished and stretched in one second (measured in Hz), which is why we hear different "high" and "low" sounds. We can say that *frequency* describes what the wave is physically doing, and *pitch* is how this frequency feels to our ears. In this book, we use the scientific term *frequency* to talk about the sound wave.

1. Ask your teacher to play this video, which helps us visualize the sound waves while playing different frequencies (http://bit.ly/1L6ZYye). Cover your ears a little if it gets irritating. (You can search "Hearing Test" on YouTube for the same video).

a. What is the lowest and highest frequency that you can hear?

> Young people can generally hear from 20 Hz to 20 kHz. Our ability to hear higher frequencies decreases as we get older.

> **Bonus note**
>
> Perhaps the whole class can close their eyes when the video starts. Then each of the students can open their eyes when they no longer hear the audio.

Unit 1 23

b. Why do you think people have different answers to Part a?

Some people can hear as low as 12 Hz and as high as 28 kHz. We are born with differences in hearing, and as we age, we lose sensitivity at the higher frequencies. There are also differences in sound equipment. We'll talk about this in the next part.

c. Some of the frequencies can only be heard with the correct sound equipment. If you can't hear some of the frequencies, is it because of the sound equipment? You need to discuss with your classmates to figure this out.

Sometimes, our computers, speakers, and headphones cannot play certain frequencies. To check if the equipment cannot play the sound, we need to work together and listen to the sound together from the same source. It also helps to have younger students listening as well. If no one can hear the sound, there is a good chance that it is due to the equipment. The equipment is either not working properly, or it was never designed to play those sound frequencies.

Bonus note

Discuss why some equipment might be designed to play a small range of sound frequencies.

2. Explain what the squiggly line in the video is showing you. What are some interesting things that you notice about it? Talk about this with your group. Share with the classroom or on the online comment section ⬉.

The squiggly lines show sound waves. Distance from left to right shows time. How high or low the line is describes how squished or stretched the air is, respectively, at a particular time. For low frequencies, we should see that the line goes up and down a few times. For high frequencies, we should see that the line goes up and down many times.

Bonus note

One might notice that the waves may appear to be traveling from one side to the other, while changing directions every 30 Hz. This has to do with the frame rate of your computer monitor.

Activity 1.2 - Octaves I

1. Let's look at how different kinds of numbers affect sound.

a. What are whole numbers and what are integers? It will be easier to remember if you describe it in your own words instead of copying from a textbook.

> Whole numbers are the counting numbers, including 0, i.e. 0, 1, 2, 3, …. Integers are the whole numbers and the negative whole numbers, i.e. 0, 1, -1, 2, -2, 3, -3, ….

b. Name three integers that are not whole numbers.

> All negative integers are not whole numbers, e.g. -1, -2, -3.

2. When the frequencies are multiples of each other by a power of two, we call them *octaves*. For example, 3 Hz and 24 Hz are octaves, because you can multiply 3 Hz by a power of two ($8 = 2^3$) to get 24 Hz. 3 Hz and 25 Hz are not octaves.

Which of the following frequencies are octaves? Write as pairs and write the lower frequency first. For example: 3 Hz & 24 Hz.

30 Hz
32.70 Hz
60 Hz
49.05 Hz
40 Hz
50 Hz
16.35 Hz

> 30 Hz & 60 Hz, and 16.35 Hz & 32.70 Hz are two pairs of octaves. None of the other pairs are multiples of powers of two of each other. Students should check that this is true!
>
> $$60 = 2 \times 30$$
> $$32.70 = 2 \times 16.35$$

3. Here are two graphs of three notes (frequencies) played together. In the first graph, all three notes are octaves of each other. In the second graph, the three notes are not octaves.

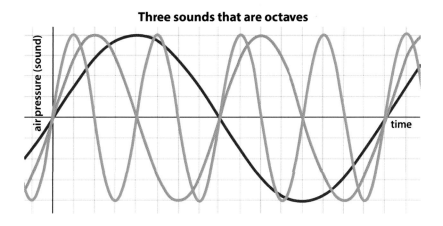

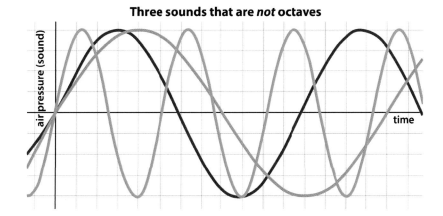

When octaves are played together, they sound really nice and sound like one note. In fact, we even give these notes the same name! We'll talk about this in Activity 1.3.

a. Compare the two graphs. What are some things that they have in common? What are some differences?

> In both graphs, the air pressure bounces up and down smoothly and predictably. In both graphs, the three curves start together in the beginning ($t = 0$s).
>
> The curves of the first graph intersect often and under convenient conditions. Within the time span displayed in the first graph, all three curves intersect three times (including at time $t = 0$s). Furthermore, there are intersections of two curves twice in this span. Conveniently, the previously mentioned points of intersection occur when the sound pressure is zero. There are many other points of intersection that are not mentioned.
>
> In the second graph, the curves intersect less often compared to the first, in fact, the intersection of all three lines rarely occurs. Intersections also occur at points all over the graph, making it hard to predict.

b. The notes in the first graph sounds nicer than the second graph. Why do you think that is?

Look at how the three waves come together in the first graph. The notes start off together (time $t = 0$ seconds), go their separate ways, and come back together twice in this time span. The notes in the first graph sound nice because the come together in a short time span and they fit nicely together. In the second graph, the notes start off together, separate, but don't come back together for quite some time. You can actually hear this "togetherness" or "lack of togetherness."

Bonus note

The "togetherness" of notes is called *consonance* and the "lack of togetherness" is called *dissonance*. We encourage the students to do more research on these terms and share their findings.

c. The notes in the first graph have the same name, but the notes in the second have different names. Why do you think this makes sense or is practical to do?

Octaves are two different notes, but they sound closer together than any other two notes played together. This is because when one note "squishes and stretches" the air once, its octave does it twice. Then both notes come together. All other whole number combinations require more "squishes and stretches" to come together. You'll see this illustrated in more detail in Activity 1.4.

If different notes come together really easily, we sometimes can't tell them apart when they're played at the same time. This is one of the main reasons why we give octaves (like one ones shown in the first graph) the same name. Also, if you're singing with the group, and the notes are too high or too low for you to sing, you can usually change all the notes down or up one octave and still sound together with everyone else.

The notes of the second graph do not come together nicely so they have different names. When two notes with different names are played together, it's easier to tell them apart.

d. Share your answers from Parts a-c with your class or in the online comment section ↖!

Bonus note

If your class is not following the pacing of the Go! Math book, you might want to skip to Activity 1.7 before returning to Activity 1.3. Activity 1.7 continues the discussion on frequency ranges, but utilizes scientific notation.

Activity 1.3 - Octaves II

Notes are specific frequencies that are named with the letters A to G (for example, B, C flat, and F sharp), which we use when playing or reading music. Recall that octaves are notes with frequencies that are powers of two multiples of each other: $2^1 f, 2^2 f, 2^3 f$, etc. So if a note has frequency f, then its octaves are $2f, 4f, 8f$, etc. We say that $2f, 4f$, and $8f$ are one, two, and three octaves higher than f.

We can also have negative powers of two: $2^{-1}f, 2^{-2}f$, and $2^{-3}f$. Therefore, $f/2, f/4$, and $f/8$ are also octaves, which are one, two, and three octaves lower than f.

Notes that are octaves apart have the same name. So for example, if F♯ (F sharp) has frequency f, then the next time that F sharp shows up is on its next octave, with frequency $2f$.

1. What frequency is 2 octaves higher than 250 Hz?

$$2^2 \times 250 \text{ Hz} = 4 \times 250 \text{ Hz}$$
$$= 1000 \text{ Hz}$$

2. What frequency is 5 octaves higher than 300 Hz?

$$2^5 \times 300 \text{ Hz} = 32 \times 300 \text{ Hz}$$
$$= 9600 \text{ Hz}$$

3. What frequency is 4 octaves lower than 4800 Hz?

$$2^{-4} \times 4800 \text{ Hz} = \tfrac{1}{16} \times 4800 \text{ Hz}$$
$$= 300 \text{ Hz}$$

4. In a 12-note scale (called a *chromatic scale*), an octave is divided into 12 frequency steps. Some notes have two names like C♯ (C sharp) and D♭ (D flat). Why do you think this is?

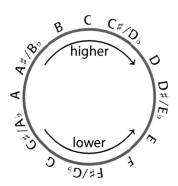

Notes with a flat ♭ or sharp ♯ sign are between two other notes that are close together. For example, C♯ (C sharp) is a note that is higher than a C but not as high as a D. D♭ (D flat) is a note that is lower than a D but not as low as C. C and D are close together but C♯ and D♭ are in between, so they're even closer together. C♯ and D♭ are so close together that we usually can't tell the difference so we consider them to be the same.

In the next activity, we will find all the notes between f and $2f$ using rational numbers. We will use a method called Pythagorean tuning. Later, we'll talk about another type of tuning.

Activity 1.4 - Notes of a Scale I

Did you notice from Activity 1.2 that the shape of a wave repeats? The *period* of a wave describes how long it takes for a wave to repeat. In a sound wave with frequency of 0.5 Hz, air squishes and stretches 0.5 times every second so we have a period of 2 seconds.

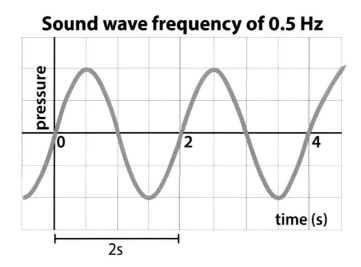

In a sound wave with frequency 1 Hz, air squishes and stretches 1 time every second so we have a period of 1 second.

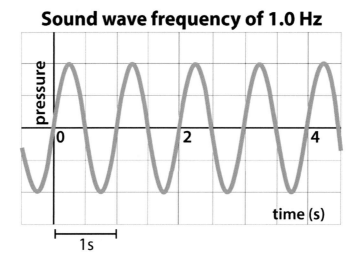

0.5 Hz and 1 Hz are octaves and sound so nice when played together that we give them the same name. Look at this graph of the two notes played together. They sound nice because *two* of one note fits nicely in *one* of the other note. In fact, notes sound nice if they are a ratio of small integers to each other, and sometimes sound awful if they are not small ratios to each other.

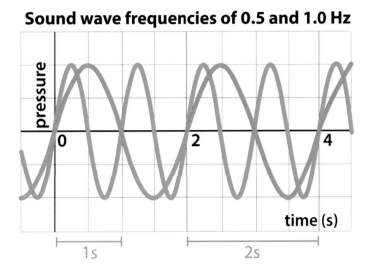

Sound wave frequencies of 0.5 and 1.0 Hz

The ratios of small integers 1/2 and 2/1 helped us find octaves. The ratios 2/3 and 3/2 will help us find all the other notes within one octave.

1. This activity will be easier if we know our powers of 2 and 3. Fill out Table 1.1.

Table 1.1: Activity 1.4 part 1. Powers of 2 and 3.

n	2^n	3^n
1	2	3
2	4	9
3	8	27
4	16	81
5	32	243
6	64	729
7	128	
8	256	
9	512	
10	1024	

2. f is the first note. To find five more we multiply a note by 3/2 to increase the frequency. Remember, we want the notes between $1f$ and $2f$ because that is the range of one octave. So if our note gets too high when we multiply by 3/2, then we have to lower the note one octave by multiplying by 1/2. Complete this guide (you can use Table 1.1, if you like):

a. **First note:** f **Multiply by 3/2:** $(3/2)f$
If note is too high, lower it one octave by multiplying by 1/2. **New note:** $(3/2)f$

b. **Previous note:** $(3/2)f$ **Multiply by 3/2:** $(9/4)f$
If note is too high, lower it one octave by multiplying by 1/2. **New note:** $(9/8)f$

c. **Previous note:** $(9/8)f$ **Multiply by 3/2:** $(27/16)f$
If note is too high, lower it one octave by multiplying by 1/2. **New note:** $(27/16)f$

d. **Previous note:** $(27/16)f$ **Multiply by 3/2:** $(81/32)f$
If note is too high, lower it one octave by multiplying by 1/2. **New note:** $(81/64)f$

e. **Previous note:** $(81/64)f$ **Multiply by 3/2:** $(243/128)f$
If note is too high, lower it one octave by multiplying by 1/2. **New note:** $(243/128)f$

3. To find the last six notes, we start with our first note f. Then we multiply a note by 2/3 to decrease the frequency. Remember, we want the notes between $1f$ and $2f$ because that is the range of one octave. So if our note gets too low when we multiply by 2/3, then we have to raise the note one octave by multiplying by 2. Complete this guide (you can use Table 1.1 if you like):

a. **First note:** f **Multiply by 2/3:** $(2/3)f$
If note is too low, raise it one octave by multiplying by 2. **New note:** $(4/3)f$

b. **Previous note:** $(4/3)f$ **Multiply by 2/3:** $(8/9)f$
If note is too low, raise it one octave by multiplying by 2. **New note:** $(16/9)f$

c. **Previous note:** $(16/9)f$ **Multiply by 2/3:** $(32/27)f$
If note is too low, raise it one octave by multiplying by 2. **New note:** $(32/27)f$

d. **Previous note:** $(32/27)f$ **Multiply by 2/3:** $(64/81)f$
If note is too low, raise it one octave by multiplying by 2. **New note:** $(128/81)f$

e. **Previous note:** $(128/81)f$ **Multiply by 2/3:** $(256/243)f$
If note is too low, raise it one octave by multiplying by 2. **New note:** $(256/243)f$

f. **Previous note:** $(256/243)f$ **Multiply by 2/3:** $(512/729)f$
If note is too low, raise it one octave by multiplying by 2. **New note:** $(1024/729)f$

Activity 1.5 - Notes of a Scale II

1. With your group, check your answers to Activity 1.4 by confirming that you found all the notes in Table 1.2 on the next page. In the examples, we gave you the first note f, the next octave $2f$, and the following four notes:

$$\frac{3}{2}f, \frac{9}{8}f, \frac{4}{3}f, \text{ and } \frac{16}{9}f$$

If you are missing some notes, work with your group to figure out where you might have made a mistake. Use teamwork!

2. What are some patterns and interesting features (or annoying ones) that you found from filling out the table? Discuss with your group and share with your classroom or on the online comment section ↖.

> We'll probably notice that when we find a note in Activity 1.4, it is difficult to predict where it'll show up on the table. Most people will have difficulty comparing fractions that do not have common denominators. Indeed we do not have common denominators in our table. The common denominator is 93312.
>
> If we look at the fractions of exponents, we see that the fractions either have the numerator as a power of two and denominator as a power of three or the numerator as a power of three and denominator as a power of two. We also see that it alternates between the two forms as the frequency increases.

3. These twelve notes are called the *chromatic scale*!

It turns out that the twelve notes of the chromatic scale have a nice pattern. If f is the frequency of a note, then the next note on the chromatic scale is either

$$\frac{2^8}{3^5}f \text{ or } \frac{3^7}{2^{11}}f$$

Look back at the Table 1.2 to check if this is true!

Table 1.2: Frequencies of a scale (Activity 1.5.1)

Order of note (lowest to highest)	Frequency (as a fraction)	Frequency (as a fraction of exponents)	Found it?
1st	f	f	✓
2nd	$\dfrac{256}{243}f$	$\dfrac{2^8}{3^5}f$	See Activity 1.4, part 3e
3rd	$\dfrac{9}{8}f$	$\dfrac{3^2}{2^3}f$	✓
4th	$\dfrac{32}{27}f$	$\dfrac{2^5}{3^3}f$	See 3c
5th	$\dfrac{81}{64}f$	$\dfrac{3^4}{2^6}f$	See 2d
6th	$\dfrac{4}{3}f$	$\dfrac{2^2}{3}f$	✓
7th	$\dfrac{1024}{729}f$	$\dfrac{2^{10}}{3^6}f$	See 3f
8th	$\dfrac{3}{2}f$	$\dfrac{3}{2}f$	✓
9th	$\dfrac{128}{81}f$	$\dfrac{2^7}{3^4}f$	See 3d
10th	$\dfrac{27}{16}f$	$\dfrac{3^3}{2^4}f$	See 2c
11th	$\dfrac{16}{9}f$	$\dfrac{2^4}{3^2}f$	✓
12th	$\dfrac{243}{128}f$	$\dfrac{3^5}{2^7}f$	See 2e
13th (next octave)	$2f$	$2f$	✓

Activity 1.6 - Notes of a Scale III

The way that you made the *chromatic scale* in Activity 1.5 is called **Pythagorean Tuning**. Instruments like the piano and keyboard use chromatic scale that is made in a slightly different way. Instead of Pythagorean tuning, these instruments use **equal temperament tuning**. In equal temperament tuning, the frequency of each note is always a number α times the last note f. So if f_1, f_2, f_3 are the first, second, and third notes, then $f_2 = \alpha f_1, f_3 = \alpha f_2 = \alpha^2 f_1$.

1. What is the difference between Pythagorean tuning and even temperament tuning?

In Pythagorean tuning, we choose one note to start with, then we repeatedly multiply it by 2/3 or 3/2 to get the other notes. We also multiply it by 2 or 1/2 to change its octave if it is too low or to high. If we sort the notes from lowest to highest, then the frequency of each note is $\dfrac{2^8}{3^5}f$ or $\dfrac{3^7}{2^{11}}f$ times that of the previous note.

In even temperament tuning, we choose one note to start with, then repeatedly multiply it by $\sqrt[12]{2}$ to get the remaining notes.

2. Write the seventh note, f_7 in terms of α and f_1.

$$\begin{aligned} f_2 &= \alpha f_1 \\ f_3 &= \alpha f_2 = \alpha^2 f_1 \\ &\cdots \\ f_7 &= \alpha f_6 = \alpha^6 f_1 \end{aligned}$$

3. The thirteenth note, f_{13} is $\alpha^{12} f_1$. Check that everyone in your group understand this.

$$\begin{aligned} f_2 &= \alpha f_1 \\ f_3 &= \alpha f_2 = \alpha^2 f_1 \\ &\cdots \\ f_{13} &= \alpha f_{12} = \alpha^{12} f_1 \end{aligned}$$

The thirteenth note is also the next octave $2f_1$. So α is the twelfth root of two.

$$\begin{aligned} f_{13} &= 2f_1 \\ \alpha^{12} f_1 &= 2f_1 \\ \div f_1 &\quad \div f_1 \\ \alpha^{12} &= 2 \\ \sqrt[12]{\alpha^{12}} &= \sqrt[12]{2} \\ \alpha &= \sqrt[12]{2} \end{aligned}$$

Compare the number $\sqrt[12]{2}$ from equal temperament with $\dfrac{2^8}{3^5}$ and $\dfrac{3^7}{2^{11}}$ from Pythagorean tuning. Write them from least to greatest. Are they close together or far apart? You may use a calculator for this.

> These numbers are very close together. By writing them as a decimal, we see that $\dfrac{2^8}{3^5} \approx 1.0535$, $\dfrac{3^7}{2^{11}} \approx 1.06787$, and $\sqrt[12]{2} \approx 1.05946$ so there are very small differences between these numbers. If we list them from least to greatest, we see that the number $\sqrt[12]{2}$ from equal temperament is in the middle of the two numbers from Pythagorean tuning. $\dfrac{2^8}{3^5} < \sqrt[12]{2} < \dfrac{3^7}{2^{11}}$.

4. Assume that the first note, f_1 is 1 Hz. Without a calculator, write the seventh note as a decimal number. Show your work and round to 2 decimal places.
Hint: In part 2 you wrote an equation for f_7 and in part 3, you have α.
Hint: $(\sqrt[b]{x})^a = x^{\frac{a}{b}}$. So for example, $\left(\sqrt[3]{12}\right)^6 = 12^{\frac{6}{3}} = 12^2 = 144$.

> The seventh note has frequency $\sqrt{2}$ Hz, which is about 1.41 Hz.
>
> $$\begin{aligned} f_7 &= \alpha^6 f_1 && \text{from part 2} \\ &= \left(\sqrt[12]{2}\right)^6 f_1 && \text{from part 3} \\ &= (2)^{\frac{6}{12}} f_1 && \text{from the hint} \\ &= (2)^{\frac{1}{2}} f_1 \\ &= (2)^{\frac{1}{2}} (1 \text{ Hz}) && \text{from the part} \\ &= (2)^{\frac{1}{2}} \text{ Hz} \\ &= \left(\sqrt{2}\right) \text{ Hz} \\ &\approx 1.41 \text{ Hz} \end{aligned}$$

5. What are *rational numbers* and *irrational numbers*? Describe in your own words.

> Rational numbers are numbers that can be written as a ratio of two integers, such as $\dfrac{1}{2}$, $2 = \dfrac{2}{1}$, and $0.\overline{1} = \dfrac{1}{9}$. Irrational numbers are numbers that cannot be written as a ratio of two integers, such as $\sqrt{2}$ and π.

Unit 1

6. Does the Pythagorean tuning use only rational numbers, only irrational numbers, or all real numbers? What about even temperament?

Pythagorean tuning uses only rational numbers, because all notes were created by repeatedly multiplying the previous note by $2, \frac{1}{2}, \frac{3}{2},$ and/or $\frac{2}{3}$. So if f is a note, all other notes are either $\frac{2^m}{3^n}f$ or $\frac{3^m}{2^n}f$ where m and n are integers. Even temperament tuning uses both rational and irrational numbers. If we start with a note that is a rational number, all octaves of this note are also rational. All other notes are irrational since they are created by multiplying $\sqrt[12]{2}$ to the previous notes. So if f is a note, all other notes have the form $\left(\sqrt[12]{2}\right)^m f$ where m is an integer.

Bonus note

Advanced students should investigate, informally prove, and share the results mentioned in the teacher notes. Here are some example questions to ask: *If one note of a Pythagorean tuned scale is a rational number, why are all of the other notes also rational? Under what conditions will a note on the even temperament scale be rational?*

Activity 1.7 - Frequency Ranges

Let's go back to the video that plays and shows different frequencies (http://bit.ly/1L6ZYye). Ask your teacher to play it in the classroom. (Cover your ears a little if it gets irritating.)

Answer the following questions in scientific notation.

1. What are the lowest and highest frequencies that you can hear from this video?

    Some of the questions in this activity have varied answers. For an example answer, let's assume that I can hear the frequencies 20 Hz to 16540 Hz. In scientific notation, this is written as 2.0×10 Hz to 1.654×10^4 Hz.

2. Use the internet to find out what's the highest frequency that a 'īlio (dog) can hear?

    According to Wikipedia, a typical 'īlio's hearing range goes up to 45 kHz. In scientific notation, this is 4.5×10^4 Hz.

3. The highest note that a 'īlio can hear is how many times higher than the highest note that you can hear?

    Note: students may have different answers. The hearing frequency range of a typical 'īlio is about 2.7 times higher than my assumed hearing range.

    $$\begin{aligned} \frac{4.5 \times 10^4 \text{ Hz}}{1.654 \times 10^4 \text{ Hz}} &= \frac{4.5}{1.654} \times \frac{10^4 \text{ Hz}}{10^4 \text{ Hz}} \\ &= \frac{4.5}{1.654} \\ &\approx 2.72 \end{aligned}$$

4. The Hawaiian spinner dolphins are called nai'a. The highest note that a nai'a can hear is about 150 thousand hertz. This is how many times higher than the highest note that you can hear?

Note: students may have different answers. The hearing frequency range of a typical nai'a is about 10 times higher than my assumed hearing range.

$$\begin{aligned}\frac{1.5 \times 10^5 \text{ Hz}}{1.654 \times 10^4 \text{ Hz}} &= \frac{1.5}{1.654} \times \frac{10^5}{10^4} \times \frac{\text{Hz}}{\text{Hz}} \\ &= \frac{1.5}{1.654} \times 10^{(5-4)} \\ &= \frac{1.5}{1.654} \times 10 \\ &\approx 0.907 \times 10 \\ &\approx 1 \times 10 \\ &= 10\end{aligned}$$

5. The speed of sound in air is about 343.2 meters per second. This is about 22.9% of the speed of sound in seawater. How does the speed of sound affect how and what you hear? How do you think this affects the lives of seals, turtles, and other animals that are always in and out of the water? Discuss with your classroom on the online comment section ↖.

With a faster speed of sound, the pressure waves reach your ears faster. Since the waves reach your ears faster, you hear more pressure waves in the same duration of time. So you actually hear a higher frequency. This is why people's voices get higher when they breathe in helium before talking. Animal that are in and out of the water frequently would benefit from being able to hear a wider frequency range. They would especially need to be able to hear the higher frequencies present in water.

Unit 2: Proportional and Nonproportional Relationships and Functions

Activity 2.1 - The Speed of Sound

Last time, we talked about the speed of sound. Let's round the speed of sound in air to 340 meters per second.

1. If we make a sound at position $y = 0$ meters at time $x = 0$ seconds, write an equation for y, where the sound wave is at time x.

$y = 340x$

2. Graph your equation from part 1. For the x axis, plot x from 0 seconds to 4 seconds. Make sure that you choose an appropriate y axis that graphs your line nicely. If the y range is too short, then the graph will fly off the top of the page right away. If the y range is too long, then the line will be really flat and may be hard to read. If it is okay with your teacher, you can use www.desmos.com/calculator.

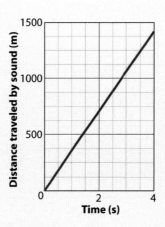

3. Suppose that you woke up late for hula practice and you're 1000 meters away from your kumu. You see that your kumu hits her ipu to the ground at time x = 0 seconds. How many seconds later do you hear the sound? Use your graph from part 2 to help you.

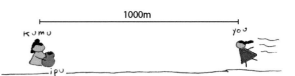

By looking at the graph from part 2, we can estimate that it will take a little less than 3 seconds for the sound of the ipu to travel 1000 m before reaching you.

4. Sometimes we hear a sound, then it bounces off of something and we hear it again. For example, if your kumu is sitting in front of big rock, 300 meters away, she will hear her ipu when she hits it. Then she hear it a second time when the sound wave travels 300 meters to the rock then bounces back another 300 meters to her. This is called an *echo*.

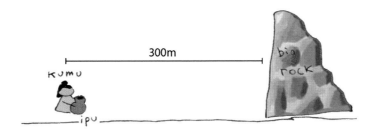

How much time passes between the first and second time that your kumu hears her ipu? Use your graph from part 2 to help you.

> The first time that kumu hears her ipu is at 0 seconds. The second time is when the sound travels 600 meters. By looking at the graph from part 2, we can estimate that it will take about 1.75 seconds for the sound to travel this distance. $1.75 - 0 = 1.75$ so 1.75 seconds pass between the first and second time that your kumu hears her ipu.

5. Suppose that you sat down between your kumu and the big rock.

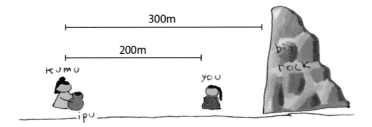

How much time passes between the first and second time that you hear the ipu? Use your graph from Part 2 to help you.

> The first time that you hear the ipu is when the sound travels 200 meters to you. According to the graph from part 2, this takes about 0.6 seconds. The second time that you hear the ipu is when the sound travels 300 meters to the big rock then another 100 meters to you. It takes about 1.2 seconds for sound to travel this distance of 400 meters. $1.2 - 0.6 = 0.6$ so 0.6 seconds pass between the first and second time that you hear the ipu.

6. Sometimes echoes can really ruin a performance. How do you think we can get rid of echoes? For example, can you just change where you are sitting? Share your ideas with your classroom on the online comment section ↖.

> If we sit right against the big rock, there will be no echo. There are many other ways to reduce echoes. We can go to a place without walls or large surfaces (on a canoe in the ocean, for example). Or we can make sure that the walls are angled and shaped in a way so that the sound waves do not head in your direction after reflecting. Or we can prevent reflections from occurring by covering walls with soft material. We encourage the students to look online for ideas and share!

Unit 2

Activity 2.2 - Song Structure

Let's learn about song structure. Most songs are broken into smaller sections such as intros, verses, and choruses (hui). The song structure describes how the sections come together to make the song. For example, many songs have the structure: intro, verse, hui, verse, hui, and hui again.

Let's take a look at the parts of our state song, Hawai'i Pono'ī, written by King David Kalākaua.

Song structure	Hawaiian	English
Verse 1	Hawai'i pono'ī	Hawai'i's own true son
	Nānā i kou mō'ī	Be loyal to your king
	Kalani ali'i	Your country's liege and lord
	Ke ali'i	The chief
Verse 2	Hawai'i pono'ī	Hawai'i's own true sons
	Nānā i nā ali'i	Look to your chiefs
	Nā pua kou muli	The flowers (children), your youngest ones
	Nā pōki'i	The young
Verse 3	Hawai'i pono'ī	Hawai'i's own true sons
	E ka lāhui ē	People of loyal heart
	'O kāu hana nui	The only duty lies
	E ui ē	List and abide
Hui	Makua lani ē	Father above us all
	Kamehameha ē	Kamehameha e
	Na kaua e pale	Who guarded in the war
	Me ka ihe	With his spear

1. Hawai'i Pono'ī usually takes 1 minute and 20 seconds to sing. This is because the song structure usually only contains the first verse, followed by the hui twice. During important Hawaiian events like the Merrie Monarch Festival, the song lasts about 2 minutes and 20 seconds. What do you think is the song structure of Hawai'i Pono'ī for important events? You can look online and/or listen to YouTube videos to see other song structures for this song.

For important events, we usually hear Hawai'i Pono'ī played with the Verse 1, Hui, Verse 2, Hui, Verse 3, Hui song structure.

2. Suppose that you and your classmates are singing a song called Mele Laina for different community events. Your class has to change the song structure to fit the different events, so your class plans to play the first few minutes of the Mele Laina, then repeat some sections x number of times until the end of the song. The hui is 40 seconds long and the verses are 1/3 minutes long.

a. Let y be the length of the performance (in minutes). If your class sings 5 minutes of Mele Laina, then repeats the hui x times, what is the equation for y?

> The length of the performance is in minutes so let's first convert the hui length from seconds to minutes.
>
> $$40 \text{ seconds} = 40 \text{ seconds} \times \frac{1 \text{ minute}}{60 \text{ seconds}} = \frac{40}{60} \text{ minute} \times \left(\frac{\text{second}}{\text{second}}\right) = \frac{2}{3} \text{ minute}$$
>
> Now, we can 2/3 minutes per repeat as the slope and 5 minutes as the y-intercept.
>
> $$y = 5 + \frac{2}{3}x$$

b. Let y be the length of the performance (in minutes). If your class sings 6 minutes of Mele Laina, then repeats the verse x times, what is the equation for y?

> We have 1/3 minutes per repeat as the slope and 6 minutes as the y-intercept.
>
> $$y = 6 + \frac{1}{3}x$$

c. Let y be the length of the performance (in minutes). If your class sings 4 minutes of Mele Laina, then repeats the entire 4 minute section x times, what is the equation for y?

> we have 4 minutes per repeat as the slope and 4 minutes as the y-intercept. Notice that we can simplify the answer with distributive property.
>
> $$\begin{aligned} y &= 4 + 4x \\ &= (4 \times 1) + (4 \times x) \\ y &= 4(1 + x) \end{aligned}$$

Unit 2

d. Let y be the length of the performance (in minutes). If your class sings 6 minutes of Mele Laina, then repeats the both the verse and the hui section x times, what is the equation for y?

For each repeat, the song gets longer by one 1/3 minute long verse and one 2/3 minute long hui. So The song gets longer by one minute per repeat. The y intercept is 6.

$$\begin{aligned} y &= 6 + \left(\frac{1}{3}x + \frac{2}{3}x\right) \\ &= 6 + \left(\frac{1}{3} + \frac{2}{3}\right)x \\ &= 6 + (1)x \\ y &= 6 + x \end{aligned}$$

3. Graph your equations from Part 2. Ask your teacher if you may use DESMOS for this part.

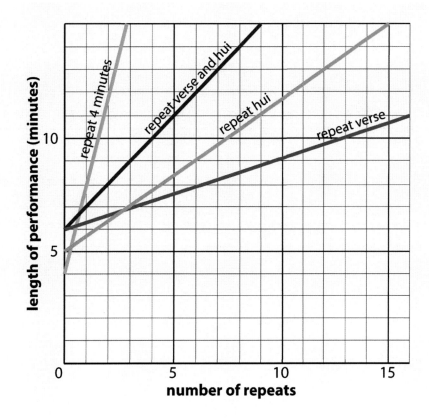

4. Another class will be dancing for 10 minutes, so your class must sing Mele Laina for 10 minutes. Which of the four song structures from Part 2 would be best for this show and why?

> There are many possible answers. What is most important is whether the student can logically justify their choice and articulate their reasoning. Here's one of many possible answers.
>
> It is a little strange to stop a song in the middle of a repeating section like a hui. So it might be a good idea to choose a song structure that reaches a song length of 10 minutes with an integer amount of repeats. This only occurs when we repeat both the verse and hui, or just the verse.

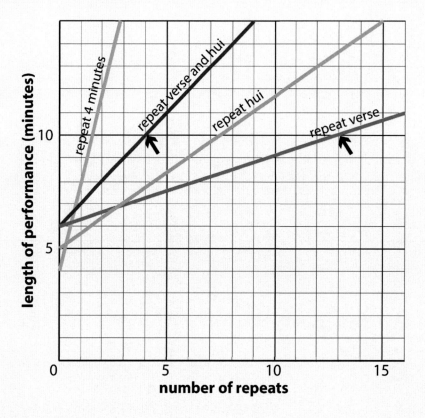

Of the two possible choices, we can reach $y = 10$ by either repeating the verse and hui four times or by repeating a verse alone 12 times. The option to repeat a verse and hui four times may be more enjoyable to some.

Activity 2.3 - String Instruments I

When you pick a 'ukulele string, it vibrates, squishing and stretching the air around it to make a sound. The frequency that the string vibrates at depends on what the string is made of, how tight the string is, and how long the vibrating part of the string is. For most songs, we only change the length of the vibrating part of the string (by holding down the string at different points).

The length of the vibrating part of the string is inversely proportional to the frequency of the sound created. This means that if the length is decreased by a rate, then the frequency is increased by the same rate.

For example, suppose a string of length L makes a sound of frequency f.

Example 1:

If we shorten the length to $(1/2)L$ by pushing the string down in the middle, then the frequency increases to $2f$.

Example 2:

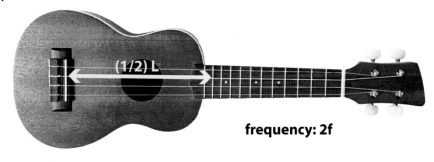

If instead, we change the length to $(2/3)L$, then the frequency changes to $(3/2)f$.

Example 3:

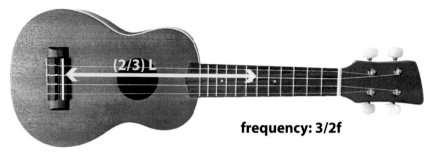

frequency: 3/2f

1. If x is the length of the vibrating string, and y is the frequency of the resulting sound, is the relationship between x and y linear or nonlinear? If it is linear, is it proportional or nonproportional? Explain. Make sure that everyone in your group is clear and agrees.

> The relationship between the length of the vibrating string (x) and the frequency of the resulting sound (y) is nonlinear. To see this we can choose three equally spaced values of x. If the relationship between x and y was linear then the corresponding values of y should be equally spaced. Let's choose $\frac{1}{4}L$, $\frac{2}{4}L$, and $\frac{3}{4}L$ as our x values. These equally spaced x values correspond to the y values $4f$, $2f$, and $\frac{4}{3}f$, which are not equally spaced.

2. The note played on the string in Example 1 is called C. What is the name of the note in Example 2? (Hint: we talked about this in Unit 1).

> The note placed in Example 2 is one octave above the note played in Example 1. Thus, the note has the same name, C.

3. The frequency of the note played in Example 1 is 261.6 Hz. What is the frequency of the note in Example 3?

> The frequency of the note played in Example 1 is $f = 261.6$ Hz. The frequency of the note played in Example 3 is $(3/2)f$. We can substitute f for 261.6 Hz to see that the note played in Example 3 has a frequency of 392.4 Hz.
>
> $$\begin{aligned}\frac{3}{2}f &= \frac{3}{2}(261.6\text{ Hz}) \\ &= 392.4\text{ Hz}\end{aligned}$$

4. Suppose that x is the length of a vibrating string, and y is the frequency of the resulting sound. Look at the following graphs. Discuss with your group if the graph correctly shows x and y. If the graph does not show the correct relationship between x and y, explain why not.

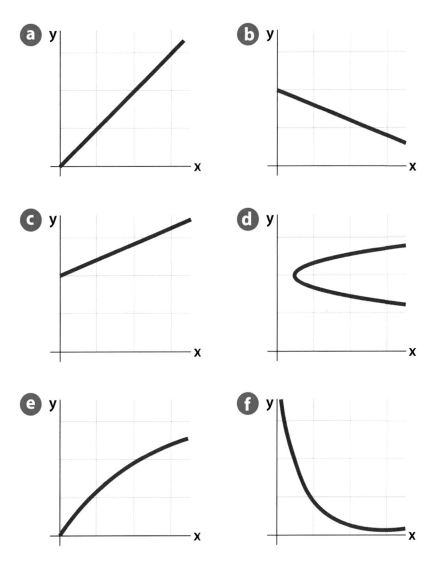

Graphs a, b, and c do not show the correct relationship between x and y. This is because, in part 1, we showed that the relationship is nonlinear. Graph d is not a function. Graph e is not correct because the frequency, y should decrease as the length x increases. Graph f is the correct graph, it is nonlinear and y decreases with increasing x. Furthermore, the y increases dramatically as x decreases closer to 0.

Activity 2.4 - String Instruments II

The strings on a 'ukulele are usually chosen and tightened so that they play the notes G (392 Hz), C (261.6 Hz), E (329.6 Hz), and A (440 Hz). As we learned earlier, we can also play other notes by shorting the length of the vibrating string. We do this by using our fingers to hold down the string at certain frets.

1. We can even find multiple ways to play the same note. For example, we can play the A note (440 Hz) by (1) picking the G string while holding the second fret, (2) picking the C string while holding fret 9, (3) picking the E string while holding fret 5, (4) and picking the A string without holding any frets.

a. Suppose that you were playing the four notes and ended up with this mapping diagram. Is "sound frequency" a function of string and fret? Why or why not? What is this mapping diagram is showing? What does it say about the 'ukulele in real life?

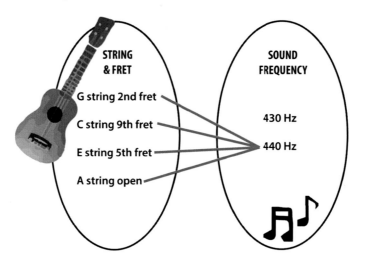

Yes, in this diagram, sound frequency is a function of string and fret. For each input (string and fret), there is one output (sound frequency) associated with it. We previously said that the four finger positions should all play the same note. This mapping diagram shows that it is true. If we played these four finger positions on a real 'ukulele and heard that they sound exactly the same, we can conclude that our 'ukulele is in tune and correctly set up.

b. Suppose that you were playing the four notes but ended up with this mapping diagram instead. Does it show a function? What does it say about your 'ukulele this time?

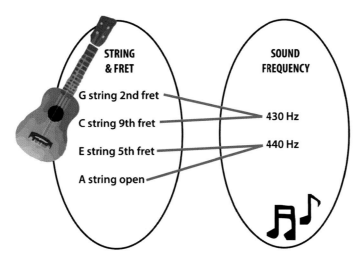

Yes, in this diagram, sound frequency is still a function of string and fret. Each input still has one output associated with it. However, we expect all four finger positions to play the same sound frequency, but it doesn't. So the 'ukulele must be out of tune, or the musician has intentionally set the 'ukulele up in an unusual way.

2. Cousins Kekoa, Kalani, Koa, Kuʻuipo, Kauʻi, and Kaimana are setting up their 'ukulele for a concert. They each play a string and a fret (left) and listen for the frequency (right).

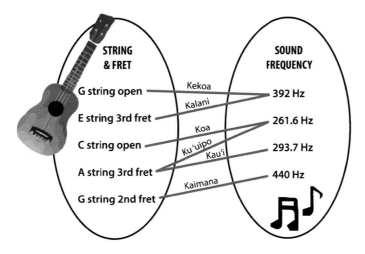

a. Which pair or pairs of cousins are playing the same note?

Kekoa and Kalani are playing the same note. Koa and Kuʻuipo are also playing the same note. It is okay for the students to include Kuʻuipo and Kauʻi, as long as they articulate their reasoning. Kauʻi has the same finger position as Kuʻuipo so he might be trying to play the same note, but he isn't.

b. Which cousin probably needs to fix the way their 'ukulele is set up? There may be multiple answers. Feel free to discuss with your group and share with the class!

> Kuʻuipo and Kauʻi are have the same finger position so they are probably trying to play the same note. One of them should probably fix their tuning to match the other. In later sections, we learn that each fret changes a note by one semitone. C is three semitones higher than A so when Kuʻuipo and Kauʻi play their note, they should sound just like their other cousin Koa. Kuʻuipo already sounds like Koa so one could argue that Kauʻi is the one who needs to fix his tuning. However, it is also possible that Kauʻi has a special tuning. So one could argue that no one needs to change their tuning. Once again, the answer is open ended. It is more important the the students think critically and articulate their reasoning.

c. This mapping diagram actually does not show a function from "string and fret" to "sound frequency". Why not?

> This does not show a function because for the input "A string 3rd fret", we have two outputs "293.7 Hz" and "440 Hz."

d. Who should stop playing in order to make this a function?

> Again, Kuʻuipo or Kauʻi needs to stop playing to make this into a mathematical function. This time, mathematically speaking, it doesn't matter which of these two cousins stop playing.

Unit 3: Solving Equations and Systems of Equations

Activity 3.1 - Stage Setup I

There are many different kinds of microphones. Choosing the right one depends mostly on what and where are you recording. If you want to record a sound coming from one direction and block out sounds from all other directions, then you might want to use a unidirectional microphone. These microphones clearly record what's right in front of it and ignores the stuff that's not in front of it.

Let's help set up the stage for a Hawaiian Choir of nine singers. On the y-axis are our three unidirectional microphones at the following points.
Mic A: (0, 4)
Mic B: (0, 8)
Mic C: (0, 13)
The Hawaiian Choir is split into three groups of three at the following points.
Group 1: (7, 4)
Group 2: (5, 8)
Group 3: (6, 12)

1. Plot these six points on a graph.

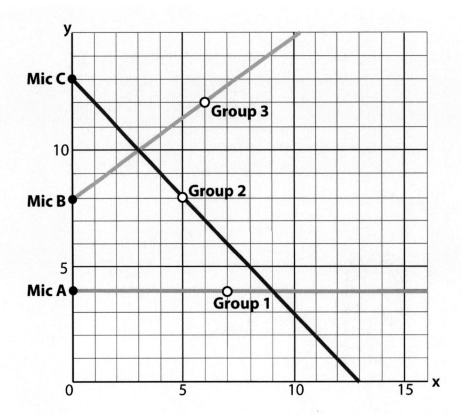

2.a. Microphone A is pointed directly at Group 1. Draw a line connecting these two points. What is the equation of the line that goes through these two points? (Use the point slope form: $y = mx + b$).

All of the microphones are on the y axis so it is easy to find the y intercept(s). The slope(s) can be found with the "rise over run" concept.

$$\begin{aligned} m &= \frac{4-4}{7-0} \\ &= \frac{0}{7} \\ &= 0 \\ b &= 4 \\ y &= mx + b \\ &= (0)x + (4) \\ y &= 4 \end{aligned}$$

b. Microphone B is pointed directly at Group 3. Draw a line connecting these two points. What is the equation of the line that goes through these two points? (Use the point slope form: $y = mx + b$).

$$\begin{aligned} m &= \frac{12-8}{6-0} \\ &= \frac{4}{6} \\ &= \frac{2}{3} \\ b &= 8 \\ y &= mx + b \\ &= \left(\frac{2}{3}\right)x + (8) \\ y &= \frac{2}{3}x + 8 \end{aligned}$$

c. Microphone C is pointed directly at Group 2. Draw a line connecting these two points. What is the equation of the line that goes through these two points? (Use the point slope form: $y = mx + b$).

$$m = \frac{8 - 13}{5 - 0}$$
$$= \frac{-5}{5}$$
$$= -1$$
$$b = 13$$
$$y = mx + b$$
$$= (-1)x + (13)$$
$$y = -x + 13$$

3. Solo! One of the singers has to perform a solo.

a. Where could they stand in order for microphones A and C to record them clearly? Use your answers from Part 2, and the substitution method.

For microphones A and C to hear the soloist clearly, they needs to stand where the microphone directional lines intersect. This is at (9, 4).

$$y = 4 \quad \text{from part 2a}$$
$$y = -x + 13 \quad \text{from part 2c}$$
$$(4) = -x + 13 \quad \text{substitution method}$$
$$+x \quad +x$$
$$x + 4 = 13$$
$$-4 \quad -4$$
$$x = 9$$

b. Where could the soloist stand in order for microphones B and C to record them clearly? Use your answers from Part 2, and the elimination method.

For microphones B and C to hear the soloist clearly, they needs to stand where the microphone directional lines intersect. This is at (3, 10).

$$y = \tfrac{2}{3}x + 8 \quad \text{from part 2b}$$
$$y = -x + 13 \quad \text{from part 2c}$$
$$y = \tfrac{2}{3}x + 8 \quad \text{elimination method}$$
$$-(y) = -(-x + 13)$$
$$y = \tfrac{2}{3}x + 8 \quad \text{elimination method}$$
$$-y = x - 13$$
$$0 = \tfrac{5}{3}x - 5$$
$$+5 \quad +5$$
$$5 = \tfrac{5}{3}x$$
$$\times \tfrac{3}{5} \quad \times \tfrac{3}{5}$$
$$3 = x$$
$$y = -(3) + 13 \quad \text{plug back into equation from part 2c}$$
$$y = 10$$

c. Does it make more sense to stand at the point from Part 3a or from Part 3b? Why?

This is open ended but most would agree that it makes more sense to stand at the point from part 3b, (3, 10), than the point from part 3a, (9, 4). This is because (9, 4) is behind the singing groups and it is very unusual for soloist to hide from the audience.

Activity 3.2 - Stage Setup II

Suppose that you set up four unidirectional microphones along the y-axis. The microphones are pointed left along the lines given by the following equations.

Mic A: $y = -2 + (2/3)x$
Mic B: $y = 2$
Mic C: $y = 6 + (2/3)x$
Mic D: $y = 8$

1. Graph the four equations for the mic directions. Label where the microphones are on the y-axis.

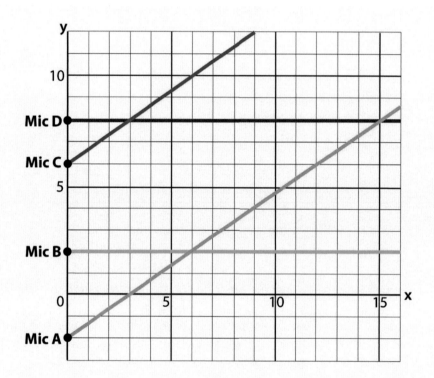

2. The soloist should stand where two microphones intersect. What is the coordinate where the soloist should stand? There is more than one possible answer, why did you choose to put the soloist here?

By looking at the graph, we can see there are three points where two microphones intersect. Students can choose any of these three points but must back up their choice with logic and articulate that logic. In these notes, we choose the microphone that is easiest to justify.

We know from life experiences that if we want to hear someone clearly, we should come reasonably close to them. The intersection of Mic C and Mic D are closest to the microphones so this point might be best suited for recording the voice of a single soloist. With the substitution method, we see that this point is at (3, 8).

$$
\begin{aligned}
y &= 8 &&\text{Mic D equation} \\
y &= 6 + \left(\frac{2}{3}\right)x &&\text{Mic C equation} \\
(8) &= 6 + \left(\frac{2}{3}\right)x &&\text{substitution method} \\
-6 & -6 \\
2 &= \left(\frac{2}{3}\right)x \\
\times \left(\frac{3}{2}\right) & \times \left(\frac{3}{2}\right) \\
3 &= x
\end{aligned}
$$

3. The backup singers should also stand where two microphones intersect, but away from the soloist. What are the coordinates where they should stand? There is more than one possible answer, why did you choose to put the backup singers here?

There are two more points left where two microphones intersect. The point that is closest to the microphones is taken by the soloist so let's choose the point that is the next closest. This is at the intersection of Mic A and Mic B. This point is at (6, 2).

$$
\begin{aligned}
y &= 2 \quad &&\text{Mic b equation} \\
y &= -2 + \left(\frac{2}{3}\right)x \quad &&\text{Mic A equation} \\
(2) &= -2 + \left(\frac{2}{3}\right)x \quad &&\text{substitution method} \\
+2 & +2 \\
4 &= \left(\frac{2}{3}\right)x \\
\times \left(\frac{3}{2}\right) & \times \left(\frac{3}{2}\right) \\
6 &= x
\end{aligned}
$$

4. There is a 'ukulele player standing at (3, 5). You have to set up one more microphone on the y-axis for their 'ukulele. This microphone is specially set up to record the sounds of a 'ukulele but microphones B and D are set up to record voices. What is the equation of the line that goes from the microphone to 'ukulele, but NEVER intersects microphones B or D?

Since the 'ukulele microphone must never intersect microphones B and D, all three microphones must have the same slope and different y intercepts. Both Mic B and Mic D have slope of $m = 0$. So our line equation becomes $y = mx + b = (0)x + b = b$. Now we need to find a y intercept, b that will cause the microphone to face the 'ukulele player at (3, 5). It is easy to see that $b = 5$ works. We can show this by replacing $x = 3$ and $y = 5$ (notice that we no longer have an x variable). So we have our equation $y = 5$.

Unit 4: Transformational Geometry

Activity 4.1 - Semitones I

A *semitone* is the difference between two notes on a scale. Many instruments can play 12 semitones within an octave. On a 'ukulele, each fret is one semitone apart. When you move your finger one fret closer to the sound hole, the frequency goes up one semitone. When you move your finger one fret further from the sound hole, the frequency goes down one semitone.

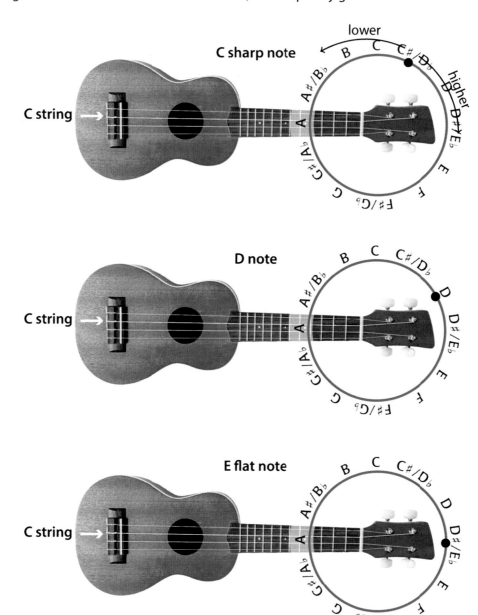

1. From the E♭ (E flat) in the last example, how many frets towards the sound hole do we have to move to play the next G note?

   G is 4 semitones higher than E♭, so we must move 4 semitones towards the sound hole.

2. What are two ways to play the C note on this string? What's the difference between these two ways?

   This is the C string so we can play the C note by playing this string open (without pressing down on any frets). There is no other way to play this exact C on this string. However, we can play all the higher octaves of C. Recall that we give all octaves the same name. Each octave has 12 semitones so we need to move our finger to the 12th fret to play the next highest C.

3. If you play the four strings (G, C, E, and A) open—without holding down any frets—then the C string will vibrate at the lowest frequency. Can a 'ukulele ever play a note lower than this note? If so, how? If not, why not?

   No, we normally cannot play a note lower than this. If we put any fingers down on the fretboard, we will shorten the vibrating string, increasing the frequency. The only way to lower the lowest note C is by changing the tuning of the strings (by loosening the string, for example).

4. How many octaves do you think can be played on a 'ukulele? Why? Share your thoughts with your class or on the online comment section ▸.

   Theoretically, there is no limit to how many octaves we can play. As long as we can shorten the vibrating strings, we can play higher frequencies. In reality, however, the strings don't vibrate well when try to play notes that are higher than what's available on the fretboard. There is also a physical limit to how short you can make the string.

Activity 4.2 - Semitones II

Use this chromatic scale circle to help you with the following parts.

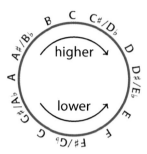

Here's our 'ukulele fretboard with a grid. Notice that the E string is on line $y = 5$. The third fret is on the line $x = 9$.

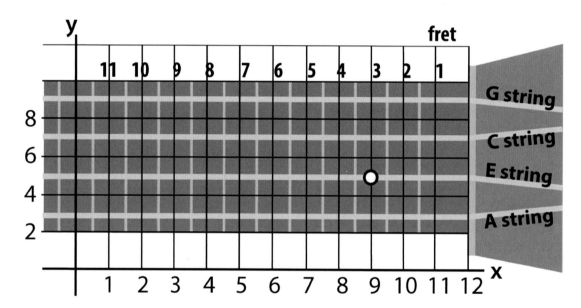

Unit 4

1. If we place a finger on (9, 5) and play the E string, then we'll hear the G note. How can you use the chromatic scale circle to show this?

   At (9, 5), our finger is on the third fret of the E string. So we know that we are playing a note that is 3 semitones higher than E. If we look at the chromatic scale circle, find the E note, and move 3 semitones clockwise, we will confirm that we are on the G note.

2. Suppose that we want to play the lowest A note on the E string. It is at $(9+\alpha, 5+\beta)$. What is the value of α and β?

   The lowest A note that is higher than E is 2 semitones higher than G. G is at (9, 5) so A must be at $(9-2, 5)$. So $\alpha = -2, \beta = 0$.

3. The lowest A note on the C string is on at $(9+\alpha, 5+\beta)$. What is the value of α and β?

   The C string is two units (one string) above the E string. So to move to the correct string, we should move from (9, 5) to $(9, 5+2)$. A is 9 semitones higher than C so we need to move to the 9th fret. We are on the 3rd fret so we need to move 6 more frets towards the sound hole from $(9, 5+2)$ to $(9-6, 5+2)$. So we have $\alpha = -6$ and $\beta = 2$.

4. Suppose that you're playing a note on (x, y). You want to play another note at $(x+\alpha, y+\beta)$. If the second note is on the same string and 3 semitones higher than the first, what is the value of α and β?

   To increase a note by 3 semitones, we need to move 3 frets toward the sound hole. Each fret is 1 unit apart so we need to move from (x, y) to $(x-3, y)$. So $\alpha = -3$.

5. Suppose that you're playing a note on (x, y). You want to play another note at $(x+\alpha, y+\beta)$. If the second note is on the same string and 2 semitones lower than the first, what is α and β?

   To decrease a note by 2 semitones, we need to move 2 frets away from the sound hole. Each fret is 1 unit apart so we need to move from (x, y) to $(x+2, y)$. So $\alpha = 2$.

Activity 4.3 - Chords

Let's learn about chords. A chord is a group of three or more notes that are heard together. Sometimes, the notes of a *chord* follow a special pattern, which has a special name.

A *major chord* is made up of three notes. The lowest note is called the *root*. The second note is called the *major third* and is four semitones higher than the root. The last note is called the *perfect fifth* and is seven semitones higher than the root. If the root has frequency f, then the perfect fifth has frequency $(3/2)f$, a note that sounds beautiful when played together with f (see Unit 1 if you don't remember why). Sometimes, we have additional notes that are octaves of other notes in this chord.

For example, we can make a chord from the root A, the major third C sharp, and the perfect fifth E. Since the root is A, and the pattern consists of a root, a major third, and a perfect fifth, this chord is called A major.

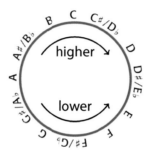

1. What are the three notes that make up a C major chord?

   To find the notes, we need to first find the root, the major third which is four semitones higher than the root, and the perfect fifth which is seven semitones higher than the root.

   Root = C
   Major third = E
   Perfect fifth = G

2. What are the three notes that make up a D major chord?

   Root = D
   Major third = F♯
   Perfect fifth = A

Here's how we can play A major on a 'ukulele. The A note is 2 semitones higher than G so we can put a finger at (10, 9). E is 4 semitones higher than C so we can put another finger at (8, 7). A is 5 semitones higher than E so we can put a finger at (7, 5). And C sharp is 4 semitones higher than A so we can put a fourth finger at (8, 3). Then we can strum the 'ukulele to play the A major chord.

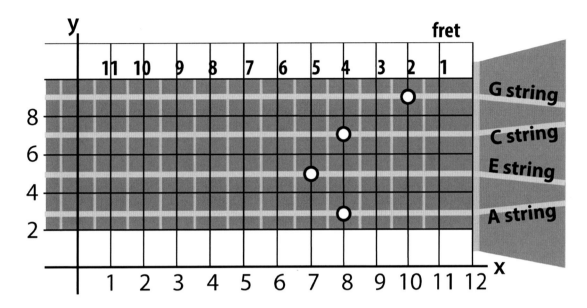

3a. Let's find other ways to play the A major chord. Use the chromatic scale circle to complete the following table.

	Root, A	Major third, C♯	Perfect fifth, E
G string	2nd fret	6th fret	9th fret
C string	9th fret	1st fret	4th fret
E string	5th fret	9th fret	open
A string	open	4th fret	7th fret

3b. Using the table from part 3a, find **two** other ways to make the A major chord. Write it as four coordinates—for example (10, 9), (8, 7), (7, 5), and (8, 3). Use 12 as the *x*-coordinate for open strings—for example, (12, 3) should represent an open A string.
Hint: You have to use all four strings to play all three notes. One string to play more than one note, but two strings can play the same note.

For the A major chord, we need the following notes.

$$\text{Root} = A$$
$$\text{Major third} = C\sharp$$
$$\text{Perfect fifth} = E$$

Let's look at the chromatic scale circle and make a table of how we can play the notes of the A major chord on each string.

	Root, A	Major third, C♯	Perfect fifth, E
G string	2nd fret	6th fret	9th fret
C string	9th fret	1st fret	4th fret
E string	5th fret	9th fret	open
A string	open	4th fret	7th fret

To make the A major chord, we can just choose to play something from each row (string). We also have to make sure that we play all three notes of the A major chord. Of course, we have four strings, which is enough to play all three notes and repeat one. For example, we can play G string, 2nd fret and A string, open to play the root A. Then we can play C string, 1st fret and E string, open to play the major third C♯ and perfect fifth E, respectively.

	Note	Finger position
Root	A	G string, 2nd fret and A string, open
Major third	C♯	C string, 1st fret
Perfect fifth	E	E string, open

Now we need to write these finger positions in terms of our coordinate system. (10, 9), (11, 7), (12, 5), (12, 3). We can find more A major chords with this method.

4. A major is made up of the notes A, C sharp, and E. Suppose that we increase each of these notes by 1 semitone to B flat, D, and F. What chord do we have now?

> If we increase the root, major third, and perfect fifth of a A major chord by one semitone, we should have the next highest major chord. If we take B♭ to be a root, we can see that D is its major third and F is its perfect fifth so we have the B♭ major chord.

5. What happens when you increase or decrease the notes of a chord by a semitone?

> If we increase or decrease all of the notes of a chord by one semitone, then we have a new chord that is one semitone higher or lower than the previous chord. Check out the last part for an example.

6. We know that one way to play A major is by putting our fingers on (10, 9), (8, 7), (7, 5), and (8, 3) and strumming. Suppose that we moved our hand to $(10+\alpha, 9+\beta)$, $(8+\alpha, 7+\beta)$, $(7+\alpha, 5+\beta)$, and $(8+\alpha, 3+\beta)$ strummed and heard F major. What are the values of α and β?

> From the second group of coordinates, we see that we are moving all of our finger positions α unit along the fretboard and β units above/below the fretboard, where α and β are the same for all fingers. It doesn't make any sense to move our fingers above or below the fretboard since there are no strings there. So $\beta = 0$. Let's take a look at the two F notes that are closest to the A note. One is four semitones lower than A and the other is 8 semitones higher. The frets on the fretboard are one unit apart so we should move all of our fingers four units away from the sound hole or eight units towards the sound hole. If we move the finger on (10, 9) four units to (14, 9), then it would be off of the fretboard. So we should move all of our coordinates eight units toward the sound hole instead. So $\alpha = -8$.

Activity 4.4 - ʻUkulele Design I

The ʻukulele comes in many different sizes. Here's a picture of a standard ʻukulele, which is also called the soprano ʻukulele.

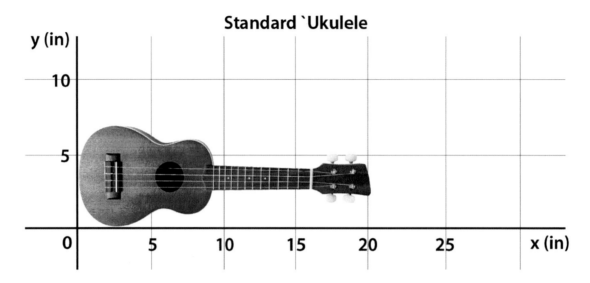

1. We can dilate the standard ʻukulele to the size of a pocket ʻukulele, which is 16 inches long.

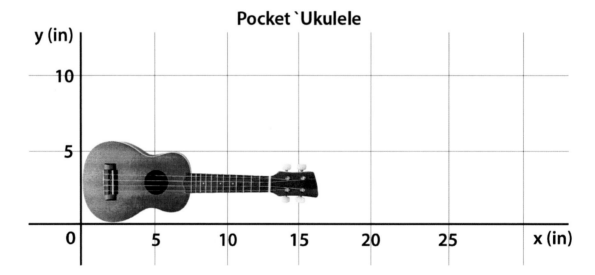

Unit 4

a. What is the scale factor of the pocket ʻukulele compared to the standard ʻukulele?
Hint: scale factor of standard ʻukulele is 1.

> We can choose a length that is on both ʻukulele and write it as a fraction. We can take any length (length of strings, length of fretboard, height of body, etc.), but it is easiest to take the length of the entire ʻukulele. The standard ʻukulele is 20 inches long and the pocket ʻukulele is 16 inches long. Writing this as a fraction gives us the following.
>
> $$\frac{16 \text{ in}}{20 \text{ in}}$$
>
> We can reduce the fraction until the side that represents the standard ʻukulele becomes 1. In this case, we will reduce until the denominator becomes 1. The numerator, which represents the relative length of the pocket ʻukulele will then become our scale factor.
>
> $$\frac{16 \text{ in}}{20 \text{ in}} = \frac{16}{20} = \frac{4}{5} = \frac{4 \div 5}{5 \div 5} = \frac{0.8}{1}$$
>
> So our scale factor is 0.8.
>
> **Bonus note**
>
> Note that we can write the fraction above, or its reciprocal. This is okay as long as we remember which one represents the standard ʻukulele. You might want to discuss why this is true with your students.
>
> $$\frac{20 \text{ in}}{16 \text{ in}} = \frac{20}{16} = \frac{5}{4} = \frac{5 \div 5}{4 \div 5} = \frac{1}{0.8}$$
>
> Whenever we divide numbers, we want to make sure that we don't divide by zero. Ask the students if we need to worry about dividing by zero in dilation problems. Why or why not? We cannot have a zero in our fraction. If we did, it would mean that we somehow stretched a point (zero length) until it had a nonzero length, and we can also shrink a nonzero length until it became a point (zero length). But *what scale factor* can we multiply to zero length to become a nonzero length? There is no such scale factor.

b. Why would someone want a pocket ʻukulele instead of a standard ʻukulele? Share your ideas with the class.

> Pocket ʻukulele is better for kids who want to learn the ʻukulele but have small hands. Pocket ʻukulele is also much more convenient to carry with when you're walking on the beach, hanging out in the mall, or riding the bus.

2. We can also dilate the standard 'ukulele to the size of a baritone 'ukulele. This baritone 'ukulele is a little bit simplified. We'll talk more about that later.

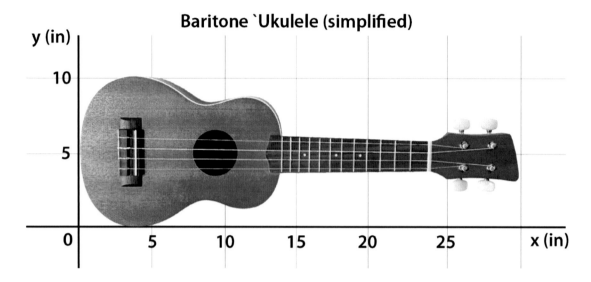

Baritone 'Ukulele (simplified)

a. What is the scale factor of the baritone 'ukulele compared to the standard 'ukulele? Hint: scale factor of standard 'ukulele is 1.

The scale factor of the baritone 'ukulele compared to the standard 'ukulele is $\frac{3}{2}$.

$$\frac{30 \text{ in}}{20 \text{ in}} = \frac{30}{20} = \frac{3}{2} = \frac{3 \div 2}{2 \div 2} = \frac{\frac{3}{2}}{1} = \frac{3}{2}$$

b. If the strings on the standard 'ukulele are L inches long, how long are the strings on the baritone 'ukulele (in terms of L)?

3/2 is the scale factor of a baritone 'ukulele compared to the standard 'ukulele. So if the strings of a standard 'ukulele is L long, then the strings of a baritone 'ukulele are $(3/2)L$ long.

c. Suppose that the strings on both the standard and the baritone 'ukulele are exactly the same, except for length. In other words, they have the same tightness and thickness, are made of the same material, etc., but they have different lengths. If you plan an open string on the standard 'ukulele and it vibrates at frequency f, what frequency will the same string vibrate at on the baritone 'ukulele?

If everything besides the length of the strings are the same, then the frequency of the string depends on the length alone. So if a string of length L vibrates at frequency f, then a string of length $(3/2)L$ should vibrate at frequency $(2/3)f$. Refer to Activity 2.3 if students need a reminder of why this is true.

d. The open strings of a standard 'ukulele play (from top to bottom) G, C, E, and A. What are the notes of the open strings of a baritone 'ukulele?
Hint: If a note has frequency f, then its *perfect fifth* has frequency $(3/2)f$. You might have to review Activity 4.3.

In the last part, we saw that a string on a baritone 'ukulele vibrates at frequency $(2/3)f$ if the same string on a standard 'ukulele vibrates at frequency f. From the hint, we also know that $(3/2)f$ is the perfect fifth of f. Consider this equation.

$$f = \frac{3}{2}\left(\frac{2}{3}f\right)$$

This and the hint tells us that the strings of the standard 'ukulele are the perfect fifths of the notes of the baritone 'ukulele. This means that the notes of the standard 'ukulele are seven semitones higher than that of the baritone 'ukulele.

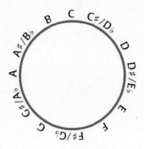

G, C, E, A are seven semitones higher than C, F, A, D, respectively. So the open strings of a baritone 'ukulele play C, F, A, and D.

e. This baritone 'ukulele is simplified. A real baritone 'ukulele has more frets. Explain why more frets are needed and support your ideas with the previous answers.

In Activity 4.1, we saw that there (theoretically) no upper limit to how high one can play on an 'ukulele. In this activity, we saw that the baritone 'ukulele can play notes that are seven semitones lower than that of a standard 'ukulele. So the baritone 'ukulele needs more frets than the standard 'ukulele, because the baritone can play all of the same notes and more.

Note that the baritone 'ukulele is designed for lower notes. Just because it can play the same notes as the standard 'ukulele, it doesn't mean that it would sound nearly as good.

Activity 4.5 - 'Ukulele Design II

The 'ukulele is traditionally a right-handed instrument—you strum and pick with your right hand. Many left-handed players just learn to play the 'ukulele right-handed. Let's look at some of the other options.

1. Some players will just rotate the right-handed 'ukulele 180° and play it upside down. (Notice the order of the strings.)

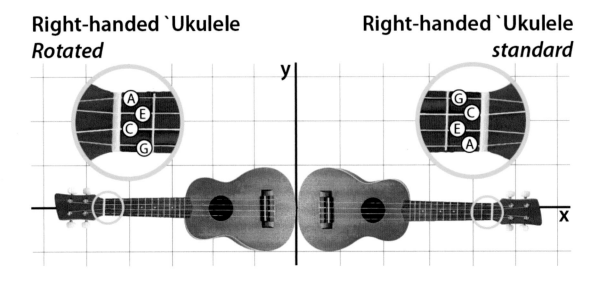

Suppose that a right-handed musician is playing a note by placing their finger at (x, y). If a left-handed musician wants to play the same note on a rotated, right-handed 'ukulele (like the one pictured), where should he/she place their finger?

> The frets have moved from the right to the left side and the strings have moved from top to bottom. So both the x and y coordinates need to be reflected from (x, y) to $(-x, -y)$

2. Some players do not want to play the right-handed 'ukulele upside down. So, in addition to rotating the right-handed 'ukulele 180°, they restring/retune the instrument so the string order is correct.

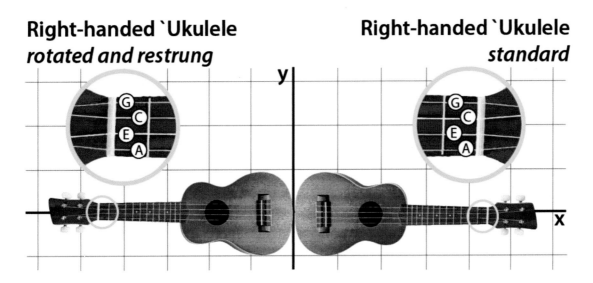

Right-handed 'Ukulele
rotated and restrung

Right-handed 'Ukulele
standard

Suppose that a right-handed musician is playing a note by placing their finger at (x, y). If a left-handed musician wants to play the same note on a rotated, restrung, right-handed 'ukulele (like the one pictured), where should he/she place their finger?

The frets have moved from the right to the left side but the string order is the same. So only the x coordinates need to be reflected from (x, y) to $(-x, y)$.

3. Some players will spend extra money to buy a left-handed 'ukulele that looks like a mirror reflection of a right-handed 'ukulele.

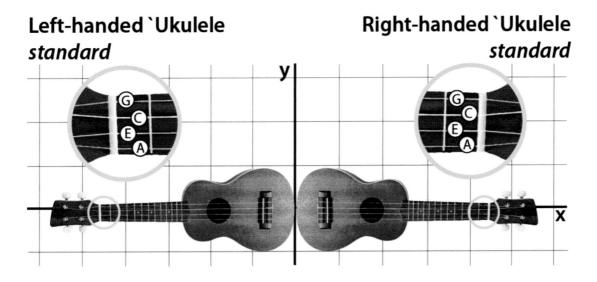

Left-handed 'Ukulele
standard

Right-handed 'Ukulele
standard

Suppose that a right-handed musician is playing a note by placing their finger at (x,y). If a left-handed musician wants to play the same note on a left-handed 'ukulele (like the one pictured), where should he/she place their finger?

> The frets have moved from the right to the left side but the string order is the same. So only the x coordinates need to be reflected from (x,y) to $(-x,y)$.

4. So a left-handed player can play a right-handed 'ukulele, a rotated right-handed 'ukulele (like in part 1), a rotated and restrung right-handed 'ukulele (like in part 2), or left-handed 'ukulele (like in part 3). Which would you choose to do? Why? Why is it a better option to you than the others?

> This is an open-ended problem. Emphasis is on forming an argument and articulating the logic behind it. Personally, I think that it is best to play a right-handed 'ukulele (in the right-handed style). This would empower a musician to pick up any available 'ukulele and play without the need to restring. This also allows an 'ukulele player to use an electronic instrument like the electric guitar that often has devices, knobs, and switches commonly placed for a right-handed player.
>
> Jimi Hendrix could play both right and left handed. When he was young, his father wanted him to play right-handed, so he learned to do that. When he played left handed, sometimes he turned his right-handed guitar around (with and without restringing). Other times, he used a left-handed guitar.
>
> Paul McCartney *never* felt comfortable playing right-handed. Although, he worked hard to learn to play right-handed, since he always played better left-handed, he just stuck with that. He used a left-handed guitar or a restrung right-handed guitar.

Unit 5: Measurement Geometry

Activity 5.1 - Sound reflections I

Sound reflects off of hard surfaces. The angle that sound approaches a surface is the same as the angle that it leaves the surface when it is reflected. For example, if you play a ipu and the sound wave hits a wall at $30°$, then it will reflect away at a $30°$ angle and continue on its way. This is called the Law of Reflection.

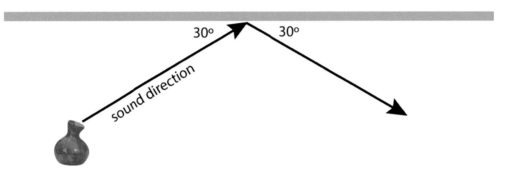

Side note: This is a simplification of how waves move. Remember that waves actually spread out in many directions instead of staying together and moving in one line. Let's just keep it simple for now. Later, we'll look at sound waves moving in different directions.

1. What are the measures of the following angles?

a. Angle a.

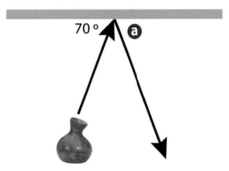

The Law of Reflections state that angle that sound approaches a surface is the same as the angle that it leaves the surface. So $a = 70°$

b. Angle b.

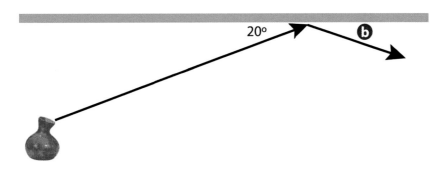

$b = 20°$

c. Angle c.

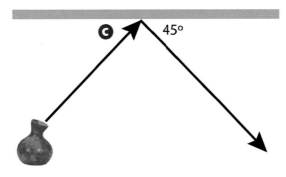

$c = 45°$

d. Angle d.

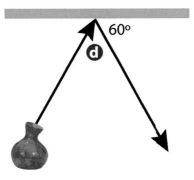

The angle of the sound leaving the wall is given to be $60°$ so by the Law of Reflection, the angle of the sound approaching the wall is also $60°$. All three angles are supplementary, meaning that they sum up to $180°$ to make up the straight wall. So $d = 60°$.

$$\begin{aligned} 60° + 60° + d &= 180° \\ 120° + d &= 180° \\ -120° & -120° \\ d &= 60° \end{aligned}$$

2. This is related to another law called Snell's Law. Look on the internet to see what Snell's Law is about.

a. What are some of the many things that Snell's Law talks about? It can get complicated so be sure to work together and share your ideas with your classmates.

> This is an exercise on researching. We do not expect the students to fully grasp the entirety of Snell's Law. We just want the students to be resourceful in learning something that they may not have heard of previously. Snell's Law discusses how light, waves, and other moving things change direction when their environment changes. For example, when light goes from air to water, when sound reaches a hard ceiling, when a pool ball hits a wall, etc.

b. Where do you see Snell's Law in real life? Share your ideas in your classroom or on the online comment section ➤.

> Some examples were mentioned in the previous teacher notes. The Hawaiians had a great intuition of how Snell's Law applies to the ocean and used it to help with fishing (throwing nets, spears, etc.).

Activity 5.2 - Sound reflections II

Recall that the Law of Reflection says that when sound approaches a hard surface and is reflected away, the angle that it approaches the surface is equal to the angle that it leaves.

1. When sound wave moves toward a corner with two walls, the sound usually reflects off of both walls before going off in some direction in the room.

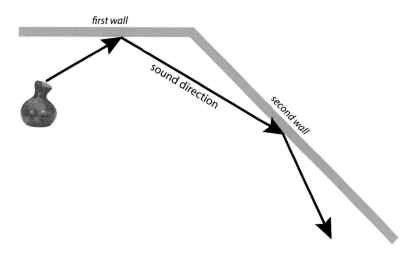

Let's take a look at what happens when the two walls are perpendicular to each other.

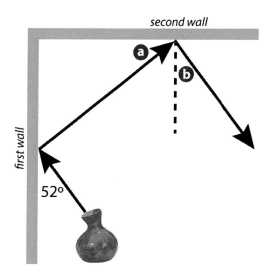

a. Find the measure of angle a. Explain your work, including when you needed to use the Law of Reflection or a math theorem.

b. The dashed line is parallel to first wall. Find the measure of angle b. Explain your work.

1. The angle of the sound leaving the first wall is $52°$ by the Law of Reflection.
2. That angle, the right angle corner, and angle a are supplementary and add up to $180°$ by the Triangle Sum Theorem. So $a = 38°$.
3. Since sound approaches the second wall at $38°$, it leaves the wall at $38°$ by the Law of Reflection.
4. Since the dash line is *parallel* to the first wall, and the first wall is *perpendicular* to the second, we know that the dash line is *perpendicular* to the second wall. We can prove this with the parallel lines theorems. Most student might not have the rigor (yet) to see that this must be proven. Feel free to point this out if you feel that your students are ready for the rigor.
5. That angle and angle b make up the angle between the second wall and the dashed line. So they are complementary. Thus $b = 52°$.

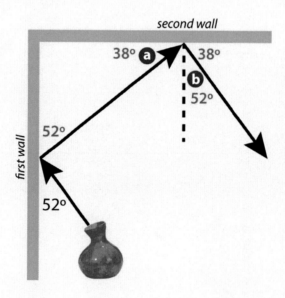

c. What can you say about the *direction of the sound wave approaching the first wall* compared to the *direction of the sound wave leaving the second wall*?

Using parallel line theorems, we see that sound wave going into perpendicular corners will leave in a parallel and opposite direction. So it will go back in the same direction it came.

d. When designing a *hale* (house or building) for listening to music, architects usually avoid having perpendicular walls in their designs. Why do you think that is?

When sound waves approach a corner with perpendicular walls, they reflect and leave the corner in the same direction that they came. In a hale with perpendicular walls, a musician, radio, or other sound source will experience an echo as a result. When sound waves are created, they will spread to the corners and come back. This echo can really ruin a musical experience.

2. Here's another 90° corner. The dashed line is parallel to the first wall. Angles a and c are congruent. Angles d and e are complementary.

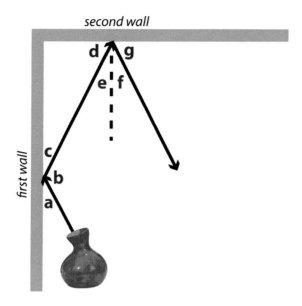

a. What are four other pairs of congruent angles? Write as pairs, for example a & c.

Angles a, c, e, and f are all congruent. Any four pairs from this list are acceptable answers. There are 6 possible answers, including the one given. We recommend having the students share how they came to their conclusions.

b. What are four other pairs of complementary angles? Write as pairs, for example d & e.

Angles d and e are complementary to every angle mentioned in part a. So any combination of d or e with a, c, e, or f are acceptable answers. There are 8 possible answers. Again, we recommend having the students share how they came to their conclusions.

3. Here's a corner that is *not* at a right angle.

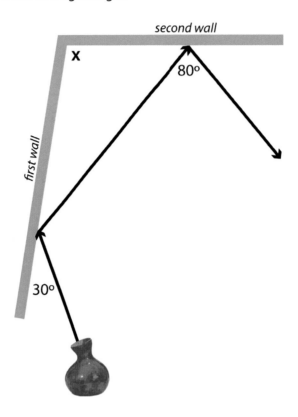

a. What is the measure of the angle of the corner, x?

1. The sound wave leaves the first wall at $30°$ by the Law of Reflection.

2. Then the sound wave approaches and leaves the second wall at the same angle by the Law of Reflection. Let's call this angle z. So z is the angle that the sound wave approaches the wall, z is also the angle that it leaves the wall, and $80°$ is the angle in between. These three angles are supplementary and sum up to $180°$. So z, the angle that the sound approaches the wall, is $50°$.

$$\begin{aligned} z + z + 80° &= 180° \\ 2z + 80° &= 180° \\ -80° &\quad -80° \\ 2z &= 100° \\ \div 2 &\quad \div 2 \\ z &= 50° \end{aligned}$$

3. Now we have that $30°$, $50°$, and x make up a triangle in the corner. We use the Triangle Sum Theorem to see that these angles add up to $180°$ so $x = 100°$.

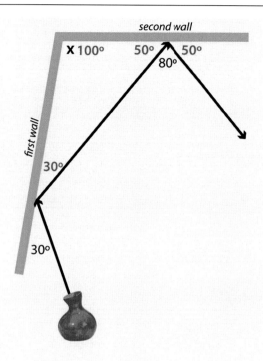

b. Think about the *direction of the sound wave approaching the first wall* compared to the *direction of the sound wave leaving the second wall.* How does this compare to the perpendicular walls in parts 1 and 2?

Since the walls are not perpendicular, the sound wave that enters the corner is not parallel to the sound wave that leaves the corner. Since the sound waves have their directions turned to other places in the room, we don't have as big of a problem with a sound bouncing back and forth in one place, creating a weird series of repeating echoes.

Activity 5.3 - Reverberations

For this activity, use 340 m/s for the speed of sound. You may also need to use the Law of Reflection at some point.

Sound waves spread out in many directions. Sometimes we need to understand the distances of things around us and how that distance affects sound.

Imagine that your hula class takes place in a valley. When kumu plays an ipu, some of the sound will go straight to you so you hear it right away. Some of the sound will actually go away from you toward mauka (the mountains). Then the sound waves will reflect off the mountains and come back to you. You might hear the reflected sound right away if the mountains are close or you might hear the sound a couple of seconds later in an echo if the mountains are far away.

1. Suppose that you're at the coordinates (0 m, 0 m), your kumu is at (40.8 m, 54.4 m), and a mountain is at (224.4 m, 299.2 m). You may use a calculator for parts a-g.

a. How far away is your kumu from you?

We can use the distance formula, which comes from the Pythagorean Theorem, to find the distance between you and kumu. This distance is 68 meters.

$$\begin{aligned} d &= \sqrt{(x_2 - x_1)^2 + (y_2 - y_1)^2} \\ &= \sqrt{(40.8 - 0)^2 + (54.4 - 0)^2} \\ &= \sqrt{40.8^2 + 54.4^2} \\ &= \sqrt{1664.64 + 2959.36} \\ &= \sqrt{4624} \\ &= 68 \end{aligned}$$

b. If kumu plays the *pū* (conch shell), how many seconds later will you first hear the sound of the pū? Round to three decimal places if needed.

We need to find the time it takes for sound to travel from the kumu to you. We can solve the system of linear equations $y = (340 \text{ m/s})t$ and $y = 68$ m to find the answer. This is equivalent to solving for t in $68 \text{ m} = (340 \text{ m/s})t$. From this, we see that it takes 0.2 seconds to first hear the sound.

$$\begin{aligned} (340 \text{ m/s})t &= 68 \text{ m} \\ \div (340 \text{ m/s}) \quad &\quad \div (340 \text{ m/s}) \\ t &= 0.2 \text{ s} \end{aligned}$$

c. How far away are the mountains from the kumu?

We use the distance formula again to show that the kumu 306 meters away from the mountains

$$\begin{aligned} d &= \sqrt{(x_2-x_1)^2+(y_2-y_1)^2} \\ &= \sqrt{(224.4-40.8)^2+(299.2-54.4)^2} \\ &= \sqrt{183.6^2+244.8^2} \\ &= \sqrt{33709+59927} \\ &= \sqrt{93636} \\ &= 306 \end{aligned}$$

d. How far away are the mountains from you?

The mountains are 374 meters away from you.

$$\begin{aligned} d &= \sqrt{(x_2-x_1)^2+(y_2-y_1)^2} \\ &= \sqrt{(224.4-0)^2+(299.2-0)^2} \\ &= \sqrt{224.4^2+299.2^2} \\ &= \sqrt{50355+89521} \\ &= \sqrt{139876} \\ &= 374 \end{aligned}$$

e. How far must the sound of the pū travel from the kumu to the mountains and then to you?

The path from the kumu to the mountains to you is 680 meters.

$$306 \text{ m} + 374 \text{ m} = 680 \text{ m}$$

f. How how many seconds after the kumu plays the pū, will you hear its sound after it reflects off of the mountains? Round to three decimal places if needed.

It takes 2 seconds for sound to travel from kumu to the mountains to you.

$$\begin{aligned} (340 \text{ m/s})t &= 680 \text{ m} \\ \div(340 \text{ m/s}) \quad &\quad \div(340 \text{ m/s}) \\ t &= 2 \text{ s} \end{aligned}$$

g. How much time passes between when you first hear the sound and when you hear its echo from the mountains?

> The first sound is heard 0.2 seconds after the kumu plays the pū. The sound is heard again 2 seconds after kumu plays the pū and the sound reflects from the mountains. The difference between these two times is 1.8 seconds.

2. If you hear a sound once, then it travels far away, reflects, and you hear it a second time, this is called an *echo*. Since sound spreads in many directions, it is more common to have sounds bouncing all over the place before reaching your ears. So we don't just hear a clear sound twice. Instead we hear a sound many times stretched out over a short duration. This is called *reverberation*. Here's a picture of sound bouncing all over the *ana* (cave) before reaching the listener.

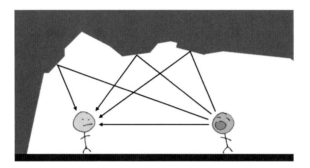

The first sound this listener hears is the sound that went straight to him/her. The last sound is the sound that bounced around the *ana* the furthest before reaching the listener. Roughly speaking, *reverberation time* is the time between when you hear the first sound and when most of the sound goes away.

Suppose that someone is singing 170 m away.

a. How long after the singer sings, does it take for the sound to first reach you? Round to three decimal places.

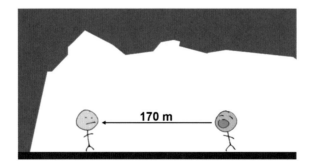

It takes 0.5 seconds for sound to travel straight from the singer to you.

$$(340 \text{ m/s})t = 170 \text{ m}$$
$$\div(340 \text{ m/s}) \quad \div(340 \text{ m/s})$$
$$t = 0.5 \text{ s}$$

b. Some of the sound does not go straight towards you, but instead bounces around first. If the slowest and last sound traveled 136 m to the ceiling before reflecting another 102 m to you, how long did it take for this sound to reach you? Round to three decimal places.

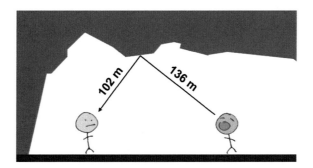

It takes 0.4 seconds for sound to travel from the singer to the ceiling and 0.3 seconds for sound to travel from the ceiling to you. This is a total of 0.7 seeconds.

$$(340 \text{ m/s})t = 136 \text{ m}$$
$$\div(340 \text{ m/s}) \quad \div(340 \text{ m/s})$$
$$t = 0.4 \text{ s}$$

$$(340 \text{ m/s})t = 102 \text{ m}$$
$$\div(340 \text{ m/s}) \quad \div(340 \text{ m/s})$$
$$t = 0.3 \text{ s}$$

c. Suppose that it's silent after the last sound wave in Part 2b reaches you. What is the reverberation time?

The reverberation time is 0.2 seconds.

$$0.7 \text{ s} - 0.5 \text{ s} = 0.2 \text{ s}$$

d. To control reverberation, we have to carefully design the angles and shape of the *ana* or where ever you like to listen to music. What is the measure of angle α? Hint: look at how the lines form a triangle. On each side of the triangle, write down how long it took sound to travel that distance.

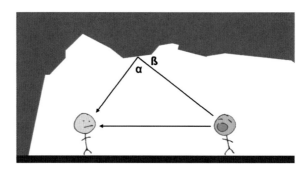

The first and last sound waves form a triangle. If we write on the sides of the triangle, the time it takes to travel that part, then we have a triangle with sides 0.3, 0.4, 0.5. This is similar triangle (scale factor 0.1) to the special 3, 4, 5 right triangle. So the angle α is $90°$.

e. Use your answer to Part d and the Law of Reflection to figure out the measure of angle β.

α, β, and a third angle are supplementary and make up the ceiling. By the Law of Reflection, β and the third angle are equal. So the angle β is $45°$.

3. Think about times where you've been in a place with a lot or very little reverberation or echo. Describe the environment. What do you think it is about those environments that give it a lot or little reverberation and echo? Share your ideas in your classroom on the online comment section ▸.

One place where I hear a lot of reverberation is in a empty hallway. This probably due to having walls close together and meeting at right angles. (Recall that we talked about perpendicular walls in Activity 5.2.) In some caves, reverberation can also be heard. In large valleys and canyons we can hear echoes. The large "walls" of a canyon and valley reflect a lot of sound and are usually far away so that there is a significant time difference between echoes. In an auditorium or concert, there usually is a pleasant amount of reverberation—not too long. These places have walls and ceilings specially designed with angles that scatter sound waves in just the right way. A room with a lot of carpeting and curtains has very little reverberation and echo. The soft material in this room absorb sound waves instead of reflecting them.

Activity 5.4 - Helmholtz Resonator

We know how the strings of a 'ukulele make sound waves. What about an ipu? A ipu works like a Helmholtz resonator, which is any instrument that is made up of a narrow tube connected to a larger hollow body. For example, a milk jug would make a great Helmholtz resonator.

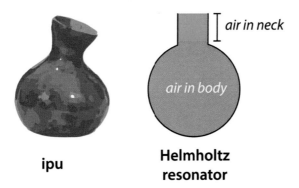

When you hit or blow across a Helmholtz resonator, the air in the neck lightly bounces in and out of the instrument. The air in the neck bounces because the air in the body can be squished and stretched. As the air in the neck bounces, it creates sound waves outside of the instrument.

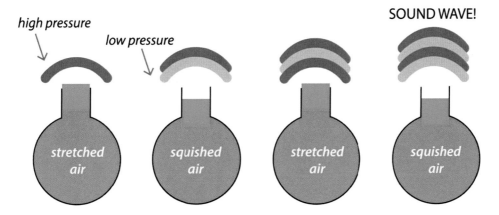

**The air in the NECK moves up and down,
while air in the BODY stretches and squishes.
The moving air in the NECK makes a sound wave outside!**

1. Suppose that you're playing several ipu in a room. They all have the same neck, but the volume of their body is V. Then the frequency that the ipu make is $f = k\sqrt{1/V}$, where k is a constant positive number.

a. If you increase the volume of the ipu body, will you get a higher or lower sound? How can you tell by looking at the formula for frequency?

> If you increase the volume V, then the denominator of the equation increases. Fractions get smaller when you increase the denominator so f gets smaller. So the sound gets lower if you increase the volume of the ipu body.

b. Rank the following ipu from *lowest* sound to *highest*. Everything expect the volumes of their bodies are the same. Assume that the body is a perfect sphere.
Ipu 1 has body radius 5 in.
Ipu 2 has body volume 900 in^3.
Ipu 3 has body diameter 13 in.
Ipu 4 has circumference of 8π inches at the widest part of the body.

> We need to order each ipu by volume. Some students will notice that volume increases with radius so it is equivalent to order these ipu by radii. They might also notice that ordering by radius is easier for all but the second ipu.
>
> | Ipu | radius (in) | volume (in^3) |
> |---|---|---|
> | Ipu 1 | 5 | 523.599 |
> | Ipu 2 | 5.989 | 900 |
> | Ipu 3 | 6.5 | 1150.347 |
> | Ipu 4 | 4 | 268.083 |
>
> Ranking from lowest to highest sound is the same as ranking from largest to smallest volume. So from low to high, we have Ipu 3, Ipu 2, Ipu 1, and Ipu 4.

c. What other objects (musical instruments or not), can be a Helmholtz resonator? Explain why and share your ideas with your classmates or on the online comment section ↖.

> In your chemistry and physics classes, you'll find a lot of examples of Helmholtz resonators. An Erlenmeyer flask and a round bottom flask are examples. At home you can find a 1-gallon milk jug and a 2-liter soda bottle that are also Helmholtz resonators. They are Helmholtz resonators because the air in the neck is in a cylindrical shape and much narrower compared to the body.

Something like a drinking straw that is closed on one end would not be a Helmholtz resonator. This is because there is no clear difference between the air in the body and the air in the neck.

Unit 6: Statistics

Activity 6.1 - Reverberation Time

Recall that sound waves spread out and bounce around. So sometimes we can still hear a sound a bit after it is no longer being made. If we hear it clearly after a short pause, then it is usually called an *echo*. If we hear it fade away–often without a pause–then it is usually a *reverberation*. As the sound waves spread, it weakens and becomes quieter. The loudness of the sound is called volume and is measured in decibels (dB). This definition of volume should not be confused with the volume that describes the amount of space in an object, which is measured in liters or cubic meters.

Suppose that you're setting up a large room for a Hawaiian music concert. To test if it is set up the way you want, you hit kālaʻau sticks together once (at time 0 ms) and use a machine to record the volume of the sound. You can watch a video about kālaʻau sticks on the book's website (https://www.stemd2.com).

1. Here is a graph of the volume in the recording from the first setup.

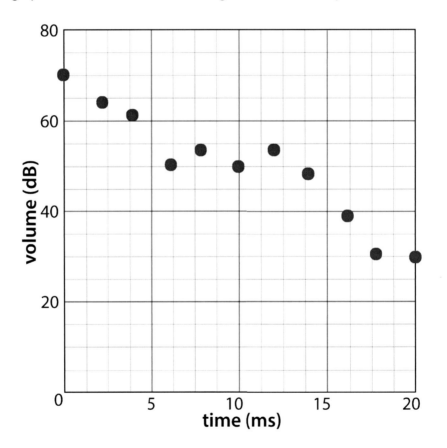

a. Does this scatter plot show a positive association, a negative association, or no association?

The sound volume typically decreases as time increases. So there is a negative association.

b. Describe the real-world reason why your answer to Part 1a makes sense.

The sound waves spread out and weaken as the air and environment absorbs the energy. The sounds that you hear later have traveled and bounced around more than the sounds that you hear right away so they're weaker.

c. Assume that there is a linear relationship shown in the graph. What is the equation for this line? Let y be loudness (volume) and t be time.

We can choose two points that look like they follow the general trend then write the line that goes through these points. The first and last point happens to look like good choices. They're at (0, 70) and (20, 30). So the equation is $y = -2x + 70$.

$$\begin{aligned} y &= mx + b \\ &= \frac{(30-70)}{(20-0)}x + (70) \\ &= -\frac{40}{20}x + 70 \\ &= -2x + 70 \end{aligned}$$

2. Here is the recording from the second setup.

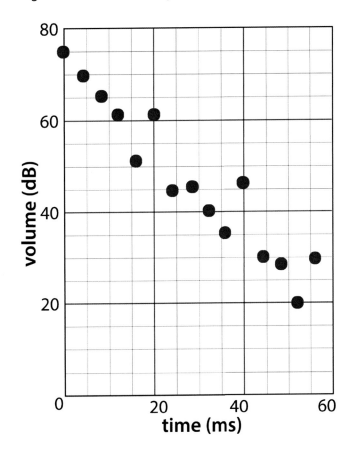

a. Does this scatter plot show a positive association, a negative association, or no association?

This also shows a negative association for the same reasons as before.

b. Describe any clusters that you see on the plot.

All but three or four points seem to fit nicely on a line. We can consider these points to be clusters. However, since the points are spread out pretty evenly, some people might not consider them to be clusters.

c. Describe some of the outliers on the plot.

The outliers are around (20, 62), (40, 47), and (57, 30). They are fairly evenly spaced out in time and a bit louder than the rest of the sounds.

d. Take a look at the outliers you found in part 2c. Why might there be outliers? Are these readings higher or lower than expected? Why might they be going higher/lower?

These outliers might be echoes. The readings are higher than expected meaning that some of the sound is bouncing around the environment strongly while most of the other sounds fade away.

e. Assume that there is a linear relationship shown in the graph. What is the equation for this line? Let y be loudness and t be time.

The points at (0, 75) and (48, 28) seem to follow the general trend of the data. We can choose to put a line through these or other similar points. This gives us the equation $y = -x + 75$.

$$\begin{aligned} y &= mx + b \\ &= \frac{(28-75)}{(48-0)}x + (75) \\ &= -\frac{47}{48}x + 75 \\ &\approx -(1)x + 75 \end{aligned}$$

Unit 6

3. *Reverberation time* is the time it takes for a sound's volume to drop 60 dB. For example, if a sound starts at 100 dB, the reverberation time is how long it takes for the sound to dampen to 40 dB.

a. What is the reverberation time of the first setup?

The equation for the first setup is $y = -2x + 70$. The y intercept tells us that at $x = 0, y = 70$. To find the reverberation time, we need to figure out when the volume y drops 60 dB to $y = 10$. The reverberation time is 30 milliseconds.

$$
\begin{aligned}
y &= -2x + 70 \\
70 - 60 &= -2x + 70 \\
10 &= -2x + 70 \\
-70 &\quad -70 \\
-60 &= -2x \\
\div(-2) &\quad \div(-2) \\
30 &= x
\end{aligned}
$$

b. What is the reverberation time of the second setup?

The reverberation time is 60 milliseconds.

$$
\begin{aligned}
y &= -x + 75 \\
75 - 60 &= -x + 75 \\
15 &= -x + 75 \\
-75 &\quad -75 \\
-60 &= -x \\
\div(-1) &\quad \div(-1) \\
60 &= x
\end{aligned}
$$

c. One of these setups results in more echoes. Which one do you think it is and why did you think it was that one?

The second setup has more echoes. This is shown in the presence of outliers. The reverberation time is also twice as long. This tells us that the setup is not very good at stopping or scattering sound waves.

d. Which of these setups would you choose for a concert and why? Discuss and share your ideas with other groups and on the on the online comment section ↖.

Most students would agree that the first setup is better for listening to music because there is less reverberation and echo. The students are free to disagree though, as long as they articulate their reasoning.

Activity 6.2 - Music Preferences I

Now that we have the room set up for the concert, let's choose the music. We can do this by asking other students what kind of music they want to hear. Here's the data we've collected by asking 210 students about their music preferences.

Music preference

		Hawaiian music	Jawaiian music	TOTAL
Age	10-12	52	78	130
	13-15	52	28	80
	TOTAL	104	106	210

1. Let's take a look at each age group separately.

Next to letters A-G, write the relative frequency compared to the total of that **row**. Here, two boxes in the first row are filled as examples. Give your answers as a percentage and round to the nearest whole number. You may use a calculator for this part.

Music preference

		Hawaiian music		Jawaiian music		TOTAL	
Age	10-12	52	40%	78	A	130	100%
	13-15	52	B	28	C	80	D
	TOTAL	104	E	106	F	210	G

Music preference

		Hawaiian music		Jawaiian music		TOTAL	
Age	10-12	52	40%	78	A 60%	130	100%
	13-15	52	B 65%	28	C 35%	80	D 100%
	TOTAL	104	E 50%	106	F 50%	210	G 100%

2. Let's take a look at each music genre separately.
Next to letters A-G, write the relative frequency compared to the total of that **column**. Here, two boxes in the Hawaiian music column are filled as an example. Give your answers as a percentage and round to the nearest whole number. You may use a calculator for this part.

Music preference

		Hawaiian music		Jawaiian music		TOTAL	
Age	10-12	52	50	78	A	130	B
	13-15	52	C	28	D	80	E
	TOTAL	104	100%	106	F	210	G

Music preference

		Hawaiian music		Jawaiian music		TOTAL	
Age	10-12	52	50	78	A 74%	130	B 62%
	13-15	52	C 50%	28	D 26%	80	E 38%
	TOTAL	104	100%	106	F 100%	210	G 100%

3.a. Explain the meaning of the percentages you wrote in Part 1. What do they tell you?

This table of percentages look at the ages separately. For a certain age, it asks: what percentage of students in this age group prefer Hawaiian or Jawaiian music?

b. Explain the meaning of the percentages you wrote in Part 2. What do they tell you?

This table of percentages look at the music genres separately. For a certain music type, it asks: what percentage of students who love this type of music are ages 10-12 or ages 13-15?

4. Of the types of songs that you're thinking of playing for the concert, what percent of them should be Hawaiian songs? What percent of them should be Hip Hop? How did you use the data from the table to help you make your decisions?

These answers are very open-ended. For example, students can look at the first table and argue that most students aged 10-12 prefer Jawaiian music, but most students aged 13-15 prefer Hawaiian music. So we should play an equal amount of each type of music. (50% Hawaiian, 50% Jawaiian). Students can also look at the second table and argue that students who like Hawaiian music come from both age groups, but the students who like Jawaiian music are mostly aged 10-12. So it might be more enjoyable to play mostly Hawaiian music since both age groups enjoy it. (70% Hawaiian, 30% Jawaiian).

Activity 6.3 - Music Preferences II

Your kumu noticed that not everyone voted. Maybe some students wanted to hear other kinds of music. To prepare for the next concert, we let students vote for "other music." Here are the results.

Music Preference

		Hawaiian	Jawaiian	Other	TOTAL
Age	Age 10-12	55	104	21	180
	Age 13-15	57	15	8	80
	TOTAL	112	119	29	260

1. Let's take a look at all the votes.

Next to the letters A-I, write the relative frequency compared to the total number of votes (260). Give your answers as a percentage and round to the nearest whole number. Three boxes are already filled. For example, we can see that 21% of all votes were from 10-12 year old students who wanted to hear Hawaiian music. You may use a calculator for this part.

Music Preference

		Hawaiian		Jawaiian		Other		TOTAL	
Age	Age 10-12	55	21%	104	**A**	21	**B**	180	**C**
	Age 13-15	57	**D**	15	**E**	8	**F**	80	31%
	TOTAL	112	**G**	119	**H**	29	**I**	260	100%

Music Preference

		Hawaiian		Jawaiian		Other		TOTAL	
Age	Age 10-12	55	21%	104	**A** 40%	21	**B** 8%	180	**C** 69%
	Age 13-15	57	**D** 22%	15	**E** 6%	8	**F** 3%	80	31%
	TOTAL	112	**G** 43%	119	**H** 46%	29	**I** 11%	260	100%

Unit 6

2. What are some conclusions or statements that you can make about the data? For example, can you say that certain groups have a stronger preference for certain kinds of music? If you cannot make any conclusions, why not?

> There are many different answers. We can make precise conclusions like *46% of the students liked Jawaiian music* and *22% of the students surveyed like Hawaiian music and are aged 13-15*. We cannot make big general conclusions like *students like Jawaiian music more than Hawaiian music*. We also have to be careful not to assume too much. For example, just because we can say that 46% of students like Jawaiian music, we cannot say that 46% of students hate Hawaiian music. The might love both types of music, but prefer to listen to Jawaiian. We only ask about what kind of music do they prefer; we do not ask about what they dislike.

3. Share your ideas with your classmates and online. Did you see any conclusions from other groups that you disagree with? How is it possible for two people to look at the same data and come up with very different conclusions? This actually happens a lot. Talk with your other classmates about how or why this is happens.

> One of the reasons that we disagree about the interpretation of the data is because we did not clearly describe where the data comes from. This is why it is very important to write a clear Methods section in a lab/science report and cite all the sources of your data. It is not a big issue here, but another major source of confusion happens when we don't choose our independent variables very well. In one example from a textbook, students are separated by whether they preferred math or science. What about students who prefer both or neither? If, for example, 10% of students prefer science, can we say that 90% do not prefer science?

Made in the USA
Las Vegas, NV
27 September 2024